FIRE ENGINES IN COLOUR

FIRE ENGINES
IN COLOUR

by
Arthur Ingram

devised and illustrated by
Denis Bishop

LONDON

BLANDFORD PRESS

First published in 1973

© 1973 Blandford Press Ltd,
167 High Holborn, London WC1V 6PH

ISBN 0 7137 0627 9

All rights reserved. No part of this book may be reproduced or transmitted in any form or by any means, electronic or mechanical, including photocopying, recording or any information storage or retrieval system, without permission in writing from the Publisher.

Colour section printed by Colour Reproductions Ltd, Billericay
Text printed and bound in Great Britain by
C. Tinling & Co. Ltd, Prescot and London

CONTENTS

Preface	7
Introduction	9
Colour Plates	21
Descriptions	117
Fire Appliance Design	217
Appliance Types	226
Bibliography	230
Index	233

PREFACE

Although this book deals with fire-fighting appliances, equipment and methods in all parts of the world and provides illustrations and descriptions of vehicles from some fifteen different countries, a certain preponderance of British information and vehicle descriptions will be noticed. This is partly an inevitable consequence of the greater availability of British information to the authors, but also reflects the prime position of Britain in most of the great epochs of fire engine history. It was the Fire of London that started it all, fire brigades first appeared in Britain and the first American steamer was the work of an Englishman, Paul Hodge. The great American and European advances of more recent times in, for instance, the building of very large capacity apparatuses and specialised airport appliances will be found to be well represented.

The authors set out to present an informative and readable volume for the general reader dealing with this specialised subject which has attracted little attention in the field of serious industrial history. For these reasons, the text has not been over-larded with technical data. Moreover, there will no doubt appear to some readers gaps and omissions according to their personal preferences. In the preparation of the book every attempt has been made to provide a balance of types, countries, manufacturers and colours, but many worthy entries have had to be passed over. We hope nevertheless that our readers will agree that we have been successful in producing an attractive and stimulating introduction to the fascinating study of fire engine history.

Our grateful thanks go to all those companies and individuals who have helped with illustrations and information on fire-fighting, and in particular to the following: Sven Bengtson, Hugh Chambers, John Christiansen, Peter Davies, John Dyson, Bryan Edwards, G. N. Georgano, Bob Graham, Niels Jansen and John Thompson. We would also like to thank the following organisations for help so freely given: the British Home Office, the London Fire Brigade, the Ministry of Defence and Old Motor Magazine.

A particular debt of gratitude is owed to G. L. Hartner who supplied much of the information and many of the illustrations concerning machines and builders on the continent of Europe.

INTRODUCTION

The story of fire-fighting runs parallel with the history of man and his efforts at science and invention. As modern civilisation has progressed, so man has invented, developed and improved the machines and methods within his grasp in order to better his environment. He has also had to turn his attention to combating the hazards and risks which have been the by-products of his continuing advance. As civilisation has developed it has brought many new ideas for us to use to our advantage, but paradoxically all advances in modern life seem to produce drawbacks which have to be overcome in order that we may secure our continued health, comfort and safety.

Every day we hear of fresh methods, read of improved processes and see new materials, some of which bring with them urgent demands for careful thought about combating the risks inherent in their use. New materials can have a high risk of combustion, a modern process can produce toxic waste and a new method can be responsible for an increase in the number of accidents if it is not properly controlled.

It is against this ever-changing background of modern development that equipment and methods have to keep pace. Although it is the interesting development of the fire engine that this book attempts to discuss we should not lose sight of the fact that no story would be possible without that intrepid band of men who throughout the world are known as – the fire fighters.

The earliest method of combating fire was simply to throw water on to the fire using anything of a convenient size that would hold water. It was also possible to beat out the flames using branches, skins, etc. No doubt for water bags or buckets made from animal skin were the easiest method of application because they were light to carry and easy to repair. It is probable that many fires once they had gained a hold on the material which was burning were just left to burn themselves out because the means of extinction available were so inadequate for a fierce blaze.

As man progressed and built more substantial dwellings so the need arose for a better method of putting water on to the flames, especially if the buildings were beyond the reach of water thrown from a bucket. So gradually woven buckets, leather buckets, metal buckets, wooden ladders, hooks, axes, metal squirts or syringes, and then force pumps, came into use.

Probably the burning building would be some distance from the water supply, so one of the first things needed was a method of transporting water from supply to fire. For many hundreds of years the hand bucket was the only means of moving water. It was slow and wasteful. If there were many people about then a chain could be formed by arranging them in a line and passing the buckets along from hand to hand. This was all right so long as there were enough of both people and buckets, but without a plentiful supply of both the fire was sure to win.

The simple type of squirt or syringe had been known since the days of the Romans, but in those countries which had never known the early civilisations or had suffered a decline in their knowledge and standards during periods of war or other turmoil, it was necessary for them to learn of such things from the word of travellers. These squirts varied in size from a small one capable of being worked by one man to much larger types which although capable of squirting a much larger amount of water with each stroke, needed two or three men to operate them. Again these squirts needed a supply of water, so a large bucket or cistern had to be brought near to the fire so that the squirt could draw water from the cistern and deliver it onto the fire. The action of these squirts was on the same principle as a garden syringe (the type without a water tank). See plate 1 for a squirt type of appliance.

Although designs of water pump are known to have existed in the early civilisations of the Greeks and Romans, their existence appears to have been forgotten because from the pages of history we read of pumps being introduced into Europe as late as 1583. This seems to have been because the writings of the early inventors and craftsmen were not translated until about that time. So the well-known system of passing buckets along a line of volunteers from source to squirt or pump was used until such time as a pump capable of drawing water from a pond was perfected.

The early pumps were designed to throw a jet of water from where they stood which was usually as close to the fire as they could safely be positioned. The pump mechanism was made to be either within or alongside the small storage reservoir or cistern which had to be constantly kept filled by bucket chain. These early pumps took various forms. Most of their inventors saw the need to provide some form of action which converted the regular rhythmic movement of mens' arm muscles into a continuous jet of water. It was therefore necessary to have two pumps working on opposite strokes or to make the pump work on both up and down strokes (double acting).

Another problem concerning the use of these early pumps was that they were quite heavy to move around, and some were mounted on a wooden base with runners similar to a sledge so that they could be dragged to the fire and the jet turned toward the blaze. Around the sixteenth century some of the heavy engines were mounted on wheels for ease of transport.

Also about this time reference is found to an engine which was in effect a large-size squirt moved about on wheels, where the water was ejected from the nozzle by screw pressure on the piston moving inside the barrel of the squirt. See plate 2 for an illustration of this type of machine.

Records from the sixteenth and seventeenth centuries show that the design of pumps was now understood in many parts of Europe and manual engines were kept by parishes, towns, estates and institutions.

It was about 1670 that Jan van der Heyden of Holland experimented with narrow strips of hide joined by sewing to form a tube. These were then joined end to end to form the first lengths of fire hose and this was placed between the pump and the metal nozzle. The fire fighter was now given much greater flexibility in his attack on the fire, and the pump need not be positioned so close to the fire and risk being burned.

As the use of leather delivery hose became more widespread it was improved by being joined in long lengths complete with metal end fittings which helped with its transport and removed the need for it to be permanently attached to the pump. It is reported that the American firm of Sellers and Pennock of Philadelphia were the first to use copper rivets in place of sewing, and this practice soon spread to other countries. Of course leather hose lasted much longer if regularly greased to maintain its flexibility and watertight properties, but this was an irksome and messy business. Jan and Nicholas van der Heyden also perfected the use of leather hose for duty at the inlet side of the pumps and thus provided suction hose to complete the operation.

Quite often fires occurred in towns which had their own water supply to houses through pipes laid under the streets and to the firemen this was a supply to be used if required. Accordingly they used to dig up the road in order to get at the water supply. The water filled the hole dug in the road and was used as a pond from which the engines were fed. Not only did this procedure play havoc with the water supply and its pipes, but it also caused a great deal of damage to the roads. In order to protect the water pipes and roads from damage in this way it was decided to fit plugs at various intervals along the pipes so that a connection could be made by which firemen could get supplies for their engines in time of need. To gain access to the water a fireman had to remove the 'fireplug' and attach a short piece of pipe in an upright direction, thus giving us the first hydrants.

In the early part of the eighteenth century some forty years after Jan van der Heyden had produced his fire engines and hose, a London button maker, Richard Newsham, patented his design of fire engine. This design, hailed as a great invention, re-introduced the long forgotten idea of having an air chamber connected to the pump so as to equalise the pressure output from the pump. This had been the discovery of Ctesibius some 2,000 years earlier but had been forgotten until 1721 when Newsham patented his design. See plate 18 for an illustration of the Newsham style of manual pump.

After the improved manual of the Newsham era it was left to other engineers and engine builders further to improve the general principles of the manual engines. This they did by arranging longer side handles which folded up for travel, by arranging for the machine to be steered and turned, and by using metal valves instead of leather ones.

The era of the improved types of large manual engines was the 1860s when good efficient engines were being produced by such firms as Merryweather, Shand Mason, Baddeley and Roberts in London. This was the time when reasonable designs of steam fire pumps were beginning to appear although we must return to 1829 to mention that Braithwaite and Ericsson produced the first steam fire engine in that year. See plate 24 for an illustration of their engine which marked the next great step forward in the history of fire-fighting.

1835 saw the design of a floating fire engine by Braithwaite although we have to wait until 1852 for the vessel to be built. Even then it was a conversion of an old hand-powered fire pump which became the first steam floating fire engine.

The first self-propelled steam fire engine was that built by Paul Rapsey Hodge in New York in 1840. Illustrations of this machine and a few other steam fire engines, including the first European self-propelled steamer, are given in plates 23 to 30.

From 1862 onwards the number of steam fire engines in use steadily grew, and by 1866 there were ninety-one horse-drawn steamers in use in Britain, while there was said to be seventeen firms producing steamers in America and a handful in Germany and France.

The horse-drawn steamer continued to capture the fire engine market during the next thirty years so far as the larger brigades were concerned, with the self-propelled steamer just beginning to get a footing at the turn of the century. A British version of a self-moving steam pump is shown in the Merryweather Fire King in plate 37, while a German version of the idea is featured in plate 39.

Many fire engine builders continued to produce horse-drawn appliances for particular markets for very many years, and the manual pump was still a good investment for a country estate or village for some time to come.

During the first ten years or so of the twentieth century there appeared a wide variety of fire appliances all aimed at taking advantage of the new inventions and designs in self-moving vehicles. These machines used an amazing collection of combinations of power for propulsion and pumping including petrol, petrol-electric, steam, electric, electric/steam, petrol/steam and electric/petrol! A selection of machines with a combination of power sources are shown in plates 39 to 46.

After about 1915 the design of fire engines settled down to petrol motor chassis of various sizes, with the smaller types mounted on high-pressure pneumatic tyres and the heavier variety on solid rubbers. The Halley machine for Glasgow shown in fig. A is typical of the British heavy motor chassis of the period, and the Dennis appliance in fig. B is that company's design around the same era.

However, not all brigades could afford the luxury of a smart modern motor appliance costing large sums of money and many continued to use steamers, manuals and horse-drawn equipment. In some instances this state of affairs was accepted by the members of the local brigade, but in other cases the firemen got fed up with being asked to use outdated equipment. It is reported that the parish council of Brockenhurst were actively considering the purchase of modern equipment at this time because there had been several recent fires in the New Forest and one had caused the total loss of the wooden horse factory!

Following on World War I the mechanisation of brigades increased with added momentum although some of the smaller towns and villages still clung to their old appliances but used a lorry hired from the local council fleet to pull old converted horse-drawn pumps. In Britain vehicle builders such as Leyland, Dennis, Merryweather, Belsize, Halley, Argyll and Tilling-Stevens were turning out machines in a variety of sizes in an effort to motorize as many brigades as possible. These machines were of fine construction and embodied the most modern equipment but were still outside the reach of some smaller authorities. Ironically some private industrial concerns could boast much better equipped brigades than the local authority!

Nevertheless the small brigades did what they could to keep pace with the latest and powerful equipment, and a wide variety of motor

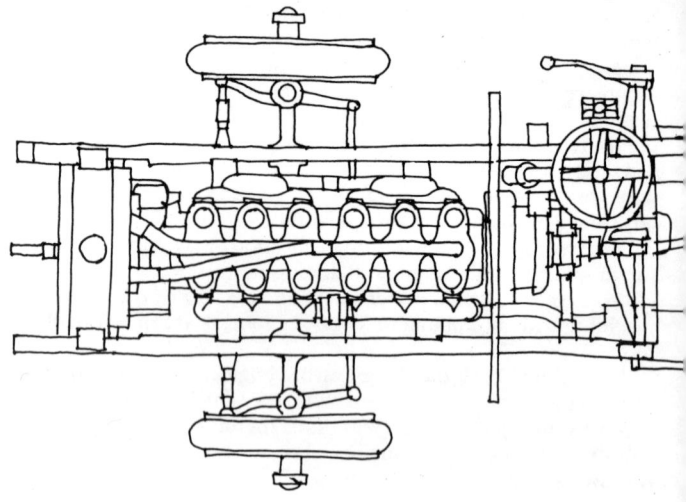

Fig. A Halley, 1915, shown without body
(UK)

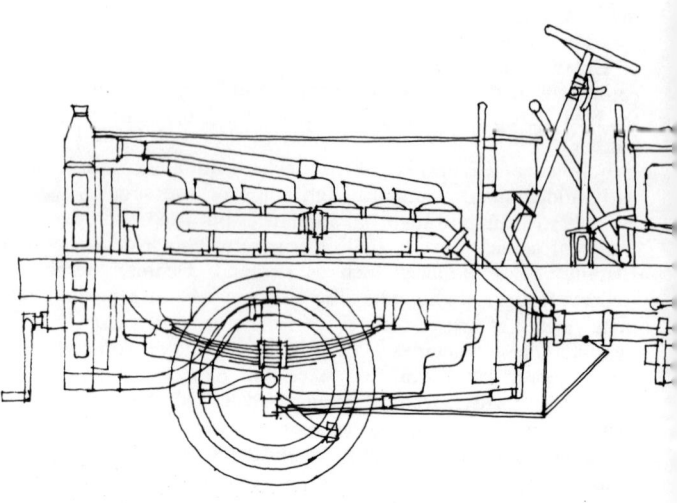

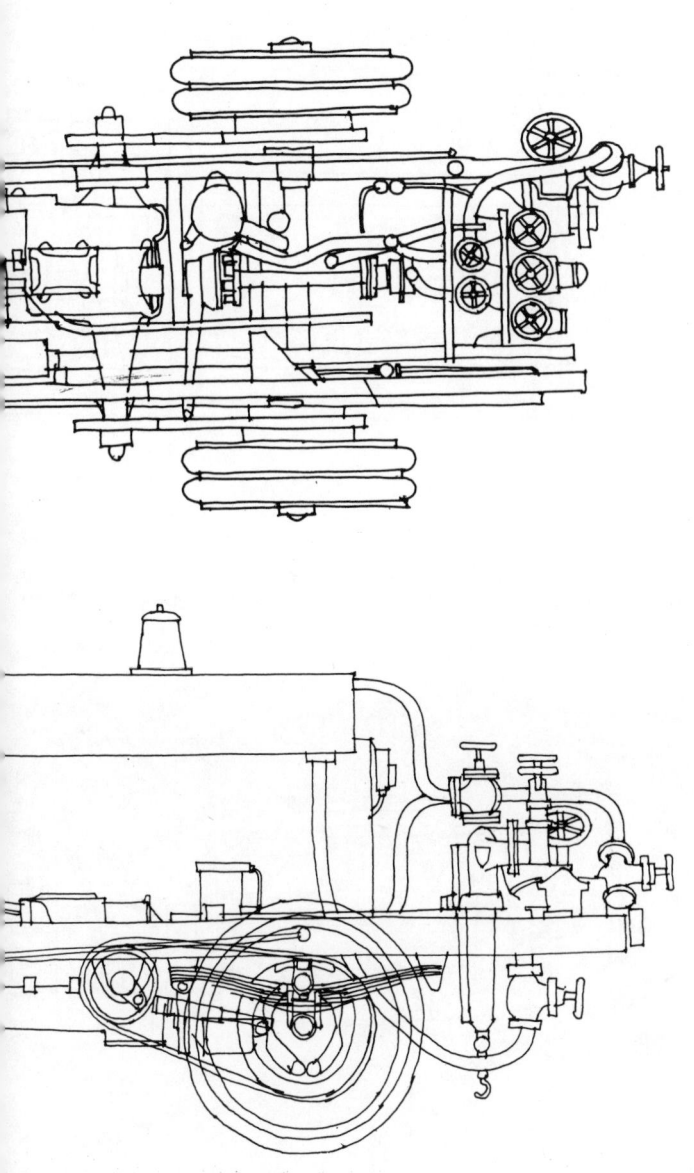

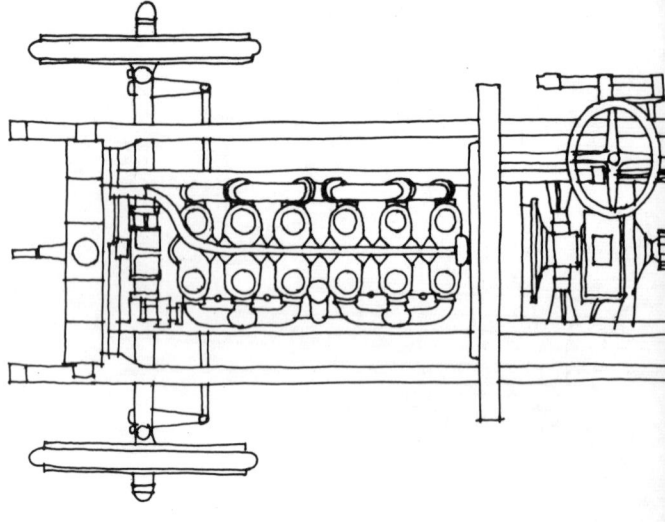

Fig. B Dennis, 1914, showing layout of equipment (UK)

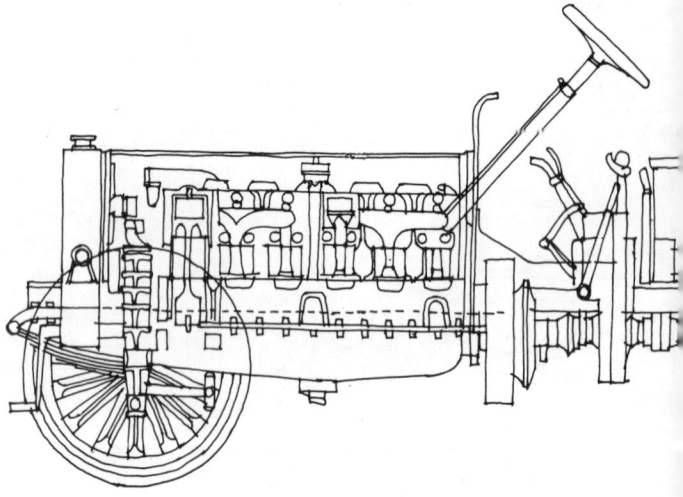

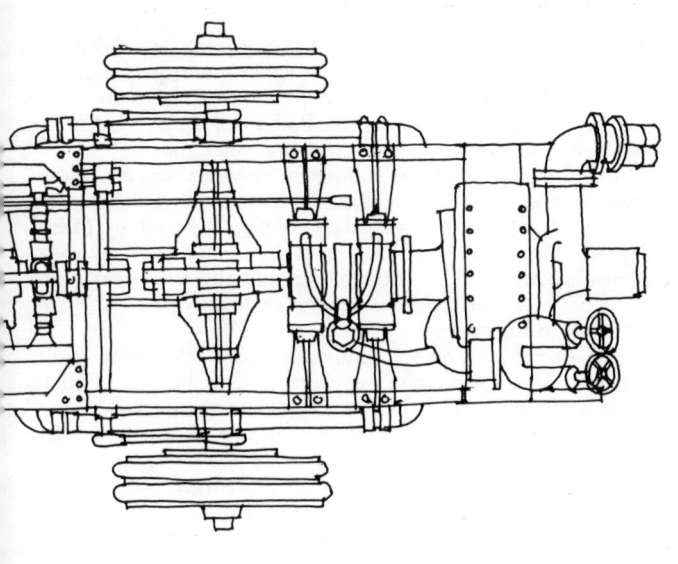

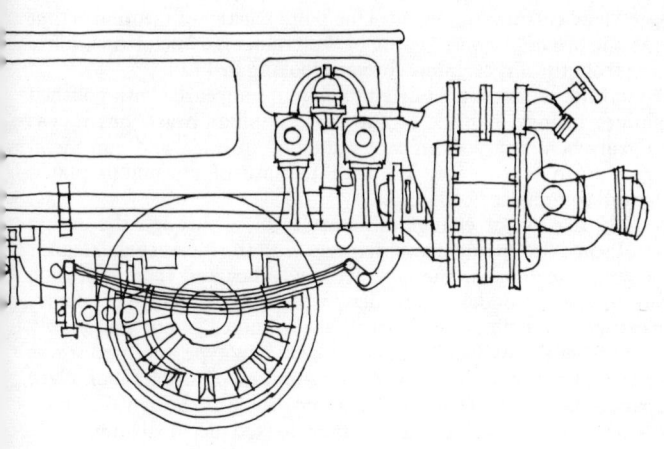

vehicles were to be seen sporting fire-fighting bodywork during the 1920s and 1930s. A list of fire engines owned by municipal brigades in 1929 reveals that in addition to the well-known names of fire engine builders there were engines on Selden, Packard, Napier, Studebaker, Daimler, GMC, Vulcan, Chambers, Star and Buick chassis.

The standard of fire-fighting equipment on water was also receiving attention and being improved. In 1926 a new fire float for the London brigade was commissioned. This was *Beta III* by Merryweather, a twin-screw vessel of 70 ft with steel hull and two petrol/paraffin engines of 110 b.h.p. each. In the same year a new vessel was put into service in Los Angeles and this vessel boasted no fewer than seven engines each of 300 b.h.p.

With the introduction of more and more motor vehicles the risk of petrol and oil fires became more apparent and in 1925 the London Fire Brigade put into service their first appliance specifically designed as a foam tender for combating this type of risk. Not only was this concern important on land but also on London's river. Following a serious fire in 1920 in which two tugs, nine barges and three fire engines were destroyed the London County Council set up a committee to study and report on the introduction of more stringent regulations for the carrying of petrol in barges.

Another innovation of the time was the fitting of safety glass for engine windscreens and the gradual adoption of giant pneumatic tyres for the heavier chassis. Other important innovations of the period were various types of foam for fighting petrol and chemical fires, the greater use of good lighting at the fire ground, and the use of oxy-acetylene cutting equipment. One point worthy of mention is that at least one fire officer was talking of the problems associated with fire fighting from the air, certainly forward looking in 1928.

Around the same time a Leyland appliance appeared with polished aluminium fittings in place of the more familiar brass, but it was many years before any significant reduction in brass and gun metal fittings was to take place, although the use of chromium plated fittings did commence about 1932.

In 1929 a wind of change did start to blow through the design offices of some fire engine manufacturers with the increasing demands for better protection for the crew from weather and road accidents. During the period under review designs called 'inside type', 'saloon', 'all-weather', and 'limousine' began to be talked about and slowly appeared from the builders' works. As each new type appeared it was under close scrutiny by many fire chiefs and much debate took place concerning the pros and cons of the design. Some of the new machines were extremely attractive while others looked decidedly ungainly.

Some officers put their faith in the inside type called 'New World', others went in for totally enclosed bodywork, while some stubbornly clung to the old Braidwood-style.

With typical British individuality fire chiefs specified the type of appliance which they personally considered best for the job and during the 1930s appliances continued to appear in a seemingly haphazard bunch of designs. This was not unusual, for there was still no standardisation of hose, couplings, fittings, pumps, etc. It took World War II and the Home Office to bring this about gradually. Just before the war started standard designs of appliance, pumping units and items of equipment were specified by the Home Office for the impending emergency.

During the 1930s France was beginning to worry about the standard of its fire force and water supplies, and in 1934 plans were laid to modernise. In Germany also there was talk of preparing for a highly skilled fire fighting force with the introduction of conscription to the service for one year, and in many quarters the topic was of possible gas attacks in a future war and the use of brigades to repel them.

By 1935 Germany had announced its intention of creating a truly modern fire service and started to standardise branches, hose, couplings, hydrants and mobile apparatus. This was the year of the International Fire Prevention and Public Security Committee Exhibition at Dresden which attracted exhibitors from twenty-six nations. To make sure the visiting fire officers were suitably impressed by the German fire service the Dresden brigade had just been re-equipped with eighteen appliances incorporating the new standards. The new fleet certainly did look impressive with front-mounted pumps capable of producing 2,400 litres per minute at 90 lbs pressure, and each vehicle powered by Maybach twelve-cylinder diesel engines of 150 h.p.

As the western world was fighting World War II so the British fire service took into use more and more of the standard appliances. The inadequacy of water supplies was to make itself felt during some of the heavy incendiary bomb attacks on large towns, and emergency supplies of water with special pipelines were set up.

At the cessation of hostilities the war-time fire fleet was dispersed to the brigades set up in 1948 and they saw many years' service, with some being extensively rebuilt and modernised for even further use. As motor manufacturers got into their stride with new designs of chassis so the brigades gradually set about replacing the outdated equipment. Many of the old pre-war names returned to the production of fire equipment while some others did not. Dennis, Merryweather, Dodge, Austin, Bedford, Commer, AEC, Ford, Land-Rover and Maudslay were there in varying numbers over the next few years with many vehicles going for export.

Almost every vehicle produced was of enclosed design and the old coach-built bodywork with ash framing and mahogany panelling gradually gave way to new methods and materials. Bodywork with plain or embossed aluminium panels, stressed panelling, metal or composite construction and later glass-reinforced plastics, were produced. The tilt cab has also come to be used in an increasing number of designs, and the custom-built machine has met with serious competition from a modified standard production chassis with short-run specialist bodywork. One interesting combination to appear in recent years has been that of ERF chassis using bodywork produced by Jennings, a firm now integrated as ERF's bodywork division. This was a company renowned as the builder of some very fine lorry chassis from 1933, although no fire engines were tackled for some thirty-odd years.

Toward the end of the 1950s the hydraulic platform made an appearance as a fire-fighting facility and this item of equipment has steadily gained favour all over the world. Not that the old turntable ladder has had its day. Far from it. With modern designs incorporating rescue cages or lifts and the ability to reach down below the horizontal plane there are still many fire chiefs who consider this type of ladder has certain advantages for rescue work particularly in confined spaces.

As we said at the beginning of this section, our fire fighters continue to keep pace with new hazards as they arise. In recent years we have witnessed developments in highly mobile cross-country appliances, in fully amphibious apparatus, in high speed emergency and rescue equipment, in fire fighting from conventional aircraft and from helicopters. A new generation of high-output aircraft crash tenders has appeared and serious thought is being given to the fighting of aircraft fires by firing ballistic missiles over the blazing aircraft. These missiles would be controlled so as to discharge their load of extinguishing material on the fire before falling to earth.

Let us continue to study the enthralling story of fire fighting of the past, and to view an exciting future safe in the knowledge that we can rely upon the continuing ingenuity of a new generation of men who are fire fighters in every sense of the word.

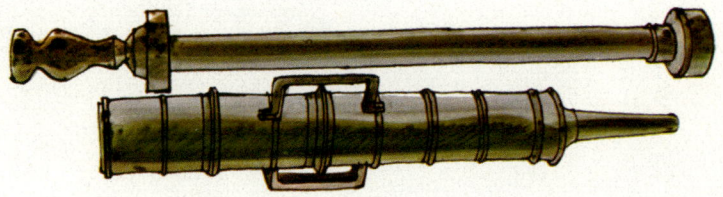

1 Seventeenth-century brass hand squirt or syringe (UK)

2 Lucar's screw-type manual pump (Germany)

Insurance Fire Marks: **4** Norwich Union Fire Insurance Society Ltd **5** Kent Fire Insurance Co. (UK)

Insurance Fire Marks: **6** Hand-In-Hand Fire and Life Insurance Society **7** Westminster Fire Office (UK)

Insurance Fire Marks: **8** Birmingham District Office **9** Queen Insurance Company (UK)

Insurance Fire Marks: **10** Bristol Crown Fire Office
11 Protector Fire Insurance Company (UK)

Insurance Fire Marks: **12** Salop Fire Office (Shropshire)   **13** Lion Fire Insurance Company (UK)

Insurance Fire Marks: **14** Sun Fire Office **15** Bristol Fire Office (UK)

16 Firemen of the Royal Exchange, London Assurance, Atlas, Sun and Westminster insurance fire brigades (UK)

17 Small tub-type appliance, seventeenth century (Belgium)

18 The Newsham design of improved manual engine, c. 1730 (UK)

19 Tub-type manual pump from Dunstable, c. 1678 (UK)

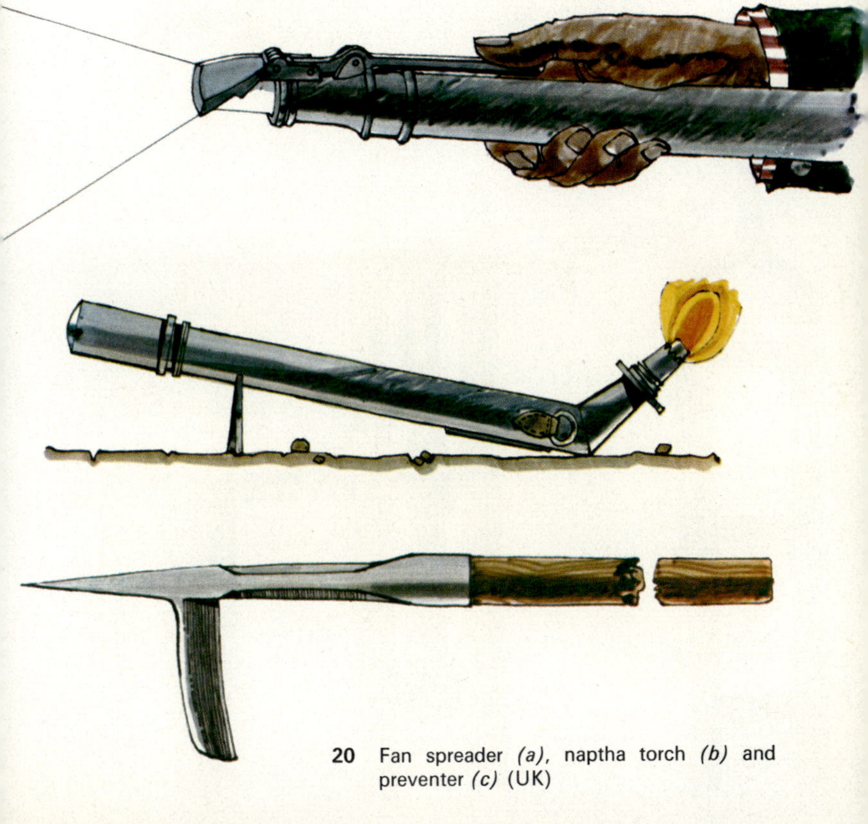

20 Fan spreader *(a)*, naptha torch *(b)* and preventer *(c)* (UK)

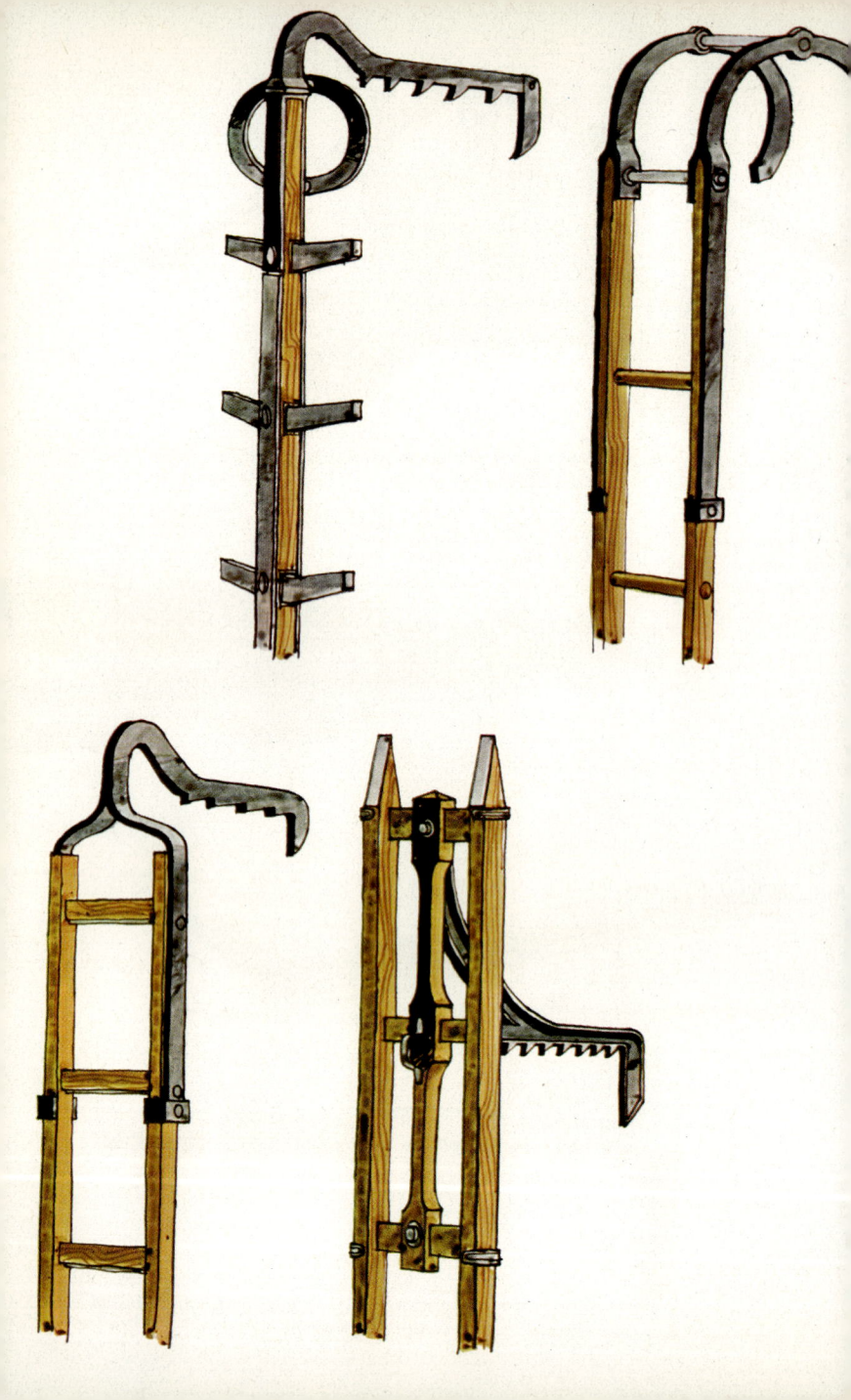

21. Designs of hook ladder from various parts of Britain and Europe

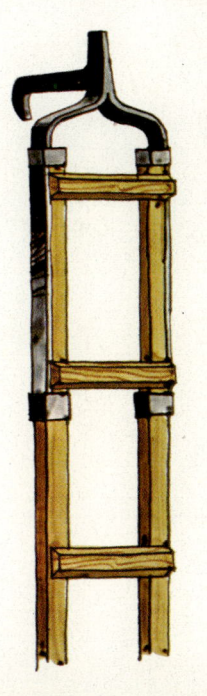

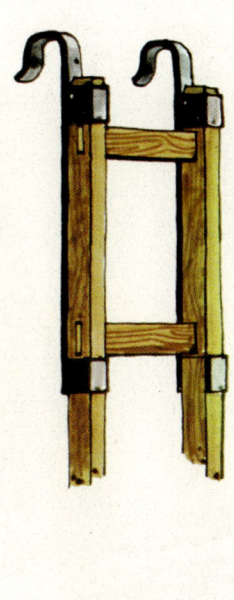

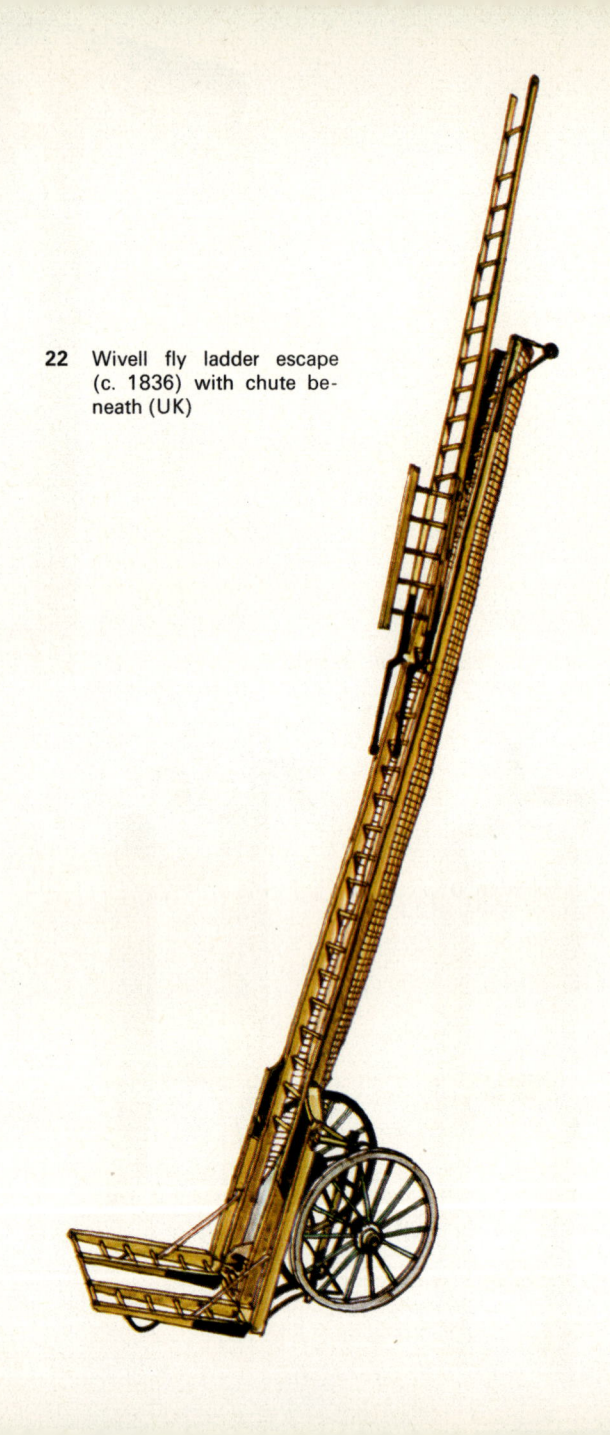

22 Wivell fly ladder escape (c. 1836) with chute beneath (UK)

23 Braithwaite's engine for Liverpool, c. 1831 (UK)

24 The epoch-making steam fire engine by Braithwaite and Ericsson, 1829 (UK)

25 Braithwaite's *Comet*, built for the King of Prussia, 1832 (UK)

26 The Hodge steamer, built in New York, 1840-1 (USA)

27 American steamer of the 1860s by Poole and Hunt (USA)

28 British steam pump of the 1860s—Merryweather's *Deluge* (UK)

29 Merryweather steam pump of 1862—the *Torrent* (UK)

30 Roberts self-propelled steamer of 1862 (UK)

31 Pirsch hook and ladder wagon, c. 1895 (USA)

32 American horse-drawn steamer, c. 1900 (USA)

33 Horse-drawn manual by Sternberg 'es Kalman, 1906 (Hungary)

34 European hybrid—Teudloff Dittrich horse-drawn water tender with petrol engine for pump, c. 1920 (Hungary)

35 Wolseley hose tender for London, 1903 (UK)

36 Merryweather motor pump (1904), the first in Britain (UK)

37 Merryweather self-propelled steamer, Fire King, 1900 (UK)

38 Cedes battery-electric turntable ladder, 1912 (UK/Germany)

39 Self-propelled steamer of Wagon und Maschinenfabrik, 1905 (Germany)

40 'Elektromobil-steam' fire engine by Braun, c. 1910 (Germany)

41 Braun electric for Charlottenburg, c. 1910 (Germany)

42 Braun-Schappler turntable ladder, c. 1910 (Germany)

43 German electric/steam pump, 1905 (Germany)

44 German electric/petrol pump, c. 1905 (Germany)

45 Braun electric/petrol for Hanover, 1911 (Germany)

46 Laurin and Klement pump, 1913 (Czecho-slovakia)

47 Merryweather pump, 1913 (UK)

48 American La France petrol/steam pumper, 1915 (USA)

49 Scania-Vabis pump, 1917 (Sweden)

50 McLaughlin airfield machine, 1918 (Canada)

51   Railroad Model T Ford, c. 1915 (USA)

52   Howe-Ford chemical apparatus, 1927 (USA)

53 Mack tractor and water tower trailer, 1925 (USA)

54 Mack tractor and aerial ladder, c. 1925 (USA)

55 Mack pumper of the 1930s (USA)

56 Leyland-BSA motorcycle combination appliance, c. 1926 (UK)

57 Clydesdale-Boyer pumper of the 1920s (USA)

58 Meray-Teudloff Dittrich motorcycle combination, c. 1928 (Hungary)

59 Commer canteen van, 1908 (UK)

60 Delahaye light-type appliance, 1926 (France)

61 Laffly Auto-Pompe Hydro-Chimique, 1926 (France)

62 Scemia sweeper/washer pump, 1927 (France)

63 Howe-Ford Model A chemical apparatus, 1928 (USA)

64 Raba-Krupp pump, c. 1916 (Hungary)

65 Raba-Austro-Fiat light pump, c. 1925 (Hungary)

66 Steyr-Rosenbauer pump, c. 1919 (Austria)

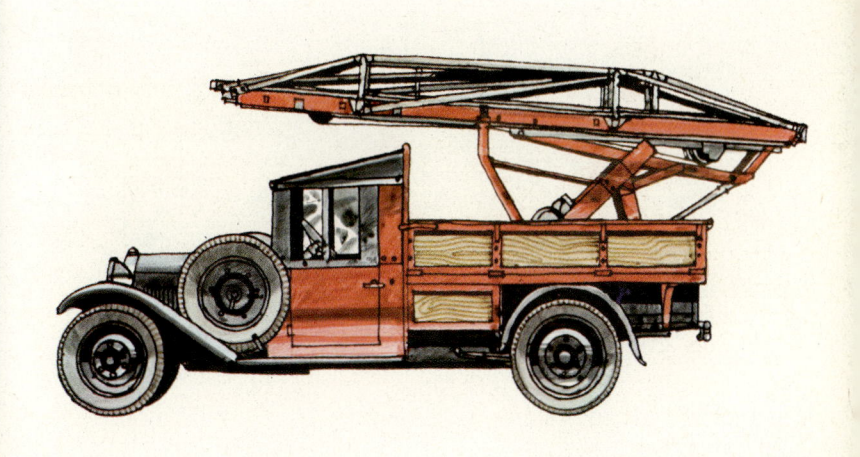

67 Steyr-Rosenbauer 14-m. ladder, c. 1919 (Austria)

68 Magomobil-Teudloff Dittrich staff car (Hungary)

69 Ahrens-Fox pumper, 1924 (USA)

70 Ahrens-Fox pumper for the New York World's Fair (USA)

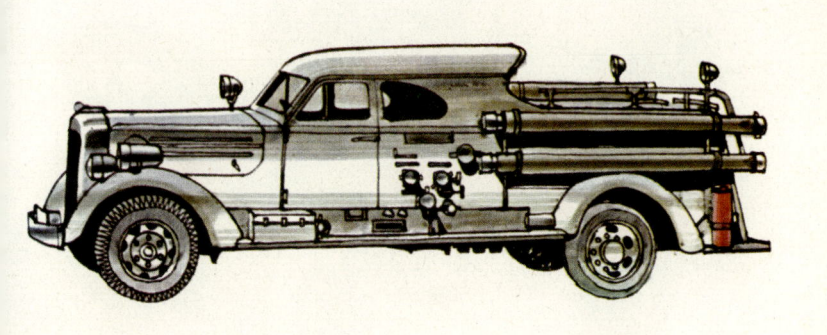

71 Ahrens-Fox 1,000-gallon pumper, 1939 (USA)

72 Rolls Royce with Braidwood-style body, 1920s (UK)

73 Fiat in service in Shanghai, 1925 (Italy)

74 Ford-Simonis pump, 1932 (UK)

75 Bedford first aid saloon tender, 1936 (UK)

76 Merryweather Hatfield pump conversion for airfields, c. 1930 (UK)

77 Morris Commercial six-wheeler, 1931 (UK)

78 Triumph car-type chemical appliance, 1932 (UK)

79 Another conversion: a 1927 Delage, with CO_2 trailer (UK)

80 Crossley streamlined 6×4 tender, 1936 (UK)

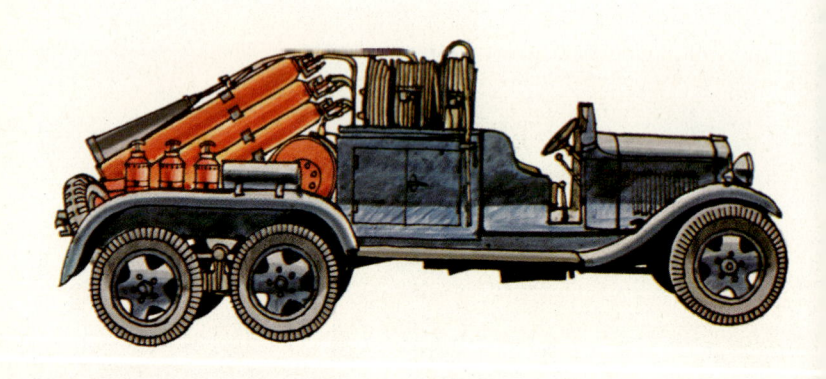

81 Ford water and CO_2 tender, 1934 (UK)

82 War-time Crossley 6×4 foam and CO_2 tender (UK)

83 Karrier Bantam war-time appliance for aerodrome protection (UK)

84 Ford WOT 1 six-wheeled foam tender, 1943 (UK)

85 Ford foam tender with folding tower and monitor, 1944 (UK)

86 Local adaptation of US Jeep as rescue tender, 1944 (USA/UK)

87 RAF conversion of captured German half-track, 1945 (Germany/UK)

88 Thornycroft-Simonis six-wheel appliance, 1931 (UK)

89 Leyland Cub six-wheeler with detachable tracks, 1933 (UK)

90 Pirsch six-wheel pump and hose apparatus (USA)

91 Leyland KZDX 2 for Speke Airport, 1939 (UK)

92 Triangel enclosed pump, 1935 (Denmark)

93 Merryweather 100-ft all-steel turntable ladder, 1933 (UK)

94 Dennis hose-layer for London, 1936 (UK)

95 Dennis canteen van for London, 1935 (UK)

96 Dennis breakdown and rescue crane, 1936 (UK)

97 Dennis Big 4 pump of the 1930s (UK)

98 Dennis Big 6 pump escape of the 1930s (UK)

99 War-time Scammell appliance on Mechanical Horse chassis (UK)

100 Ford V8 pump water tender, 1936 (Rumania)

101 Ford 7V Home Office appliance, 1939 (UK)

102 War-time Ford V8 towing unit (UK)

103 War-time Dennis–JAP light pump (UK)

104 War-time Dennis medium trailer pump (UK)

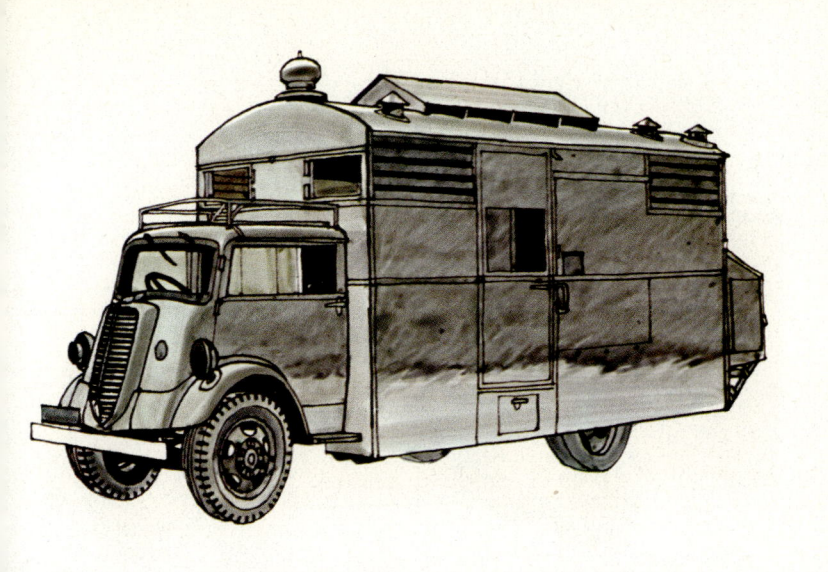

105 Ford 7V emergency kitchen, 1941 (UK)

106 War-time emergency canteen on Humber car chassis (UK)

107 War-time Ford 7V control van for London (UK)

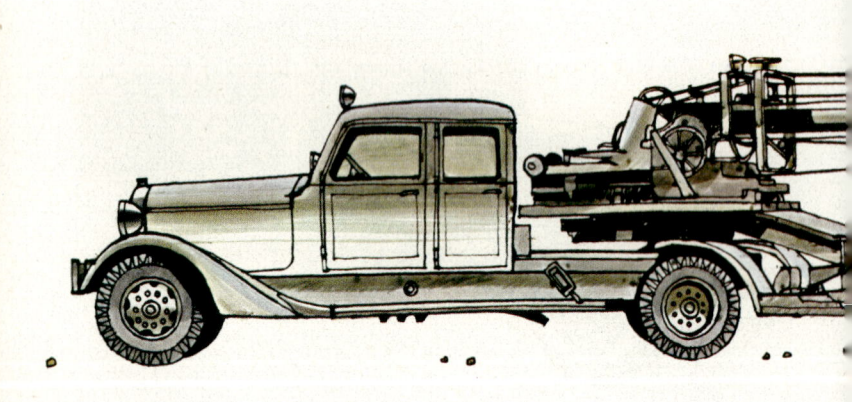

108 American La France articulated water tower, 1948 (USA)

109 Bedford-Pyrene airfield crash tender, 1949 (UK)

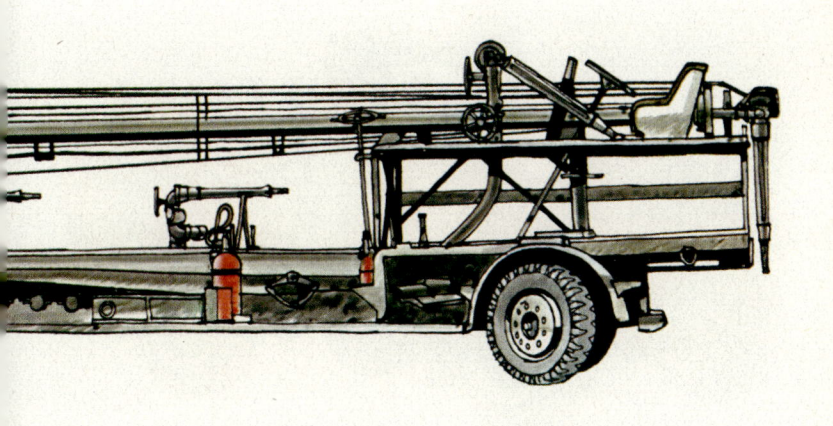

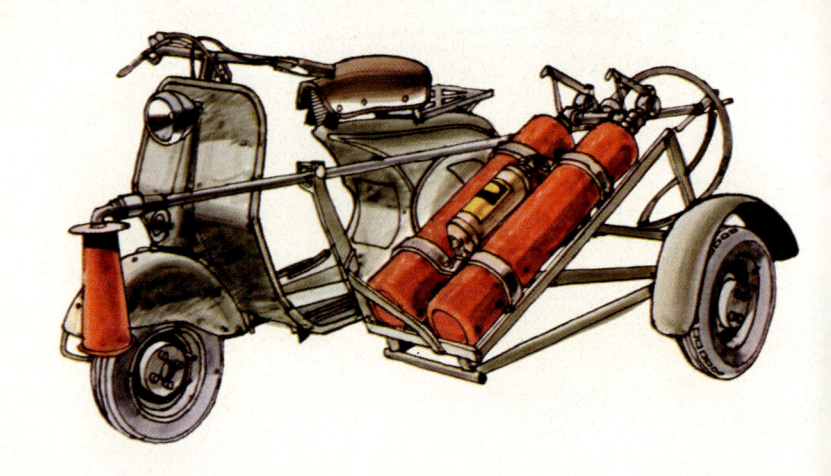

110 Vespa-Pyrene aircraft engine fire appliance (UK)

111 Morris Commercial appliance used at RAF air displays (UK)

112 Dodge-Pyrene CO_2 tender of the early post-war era (UK)

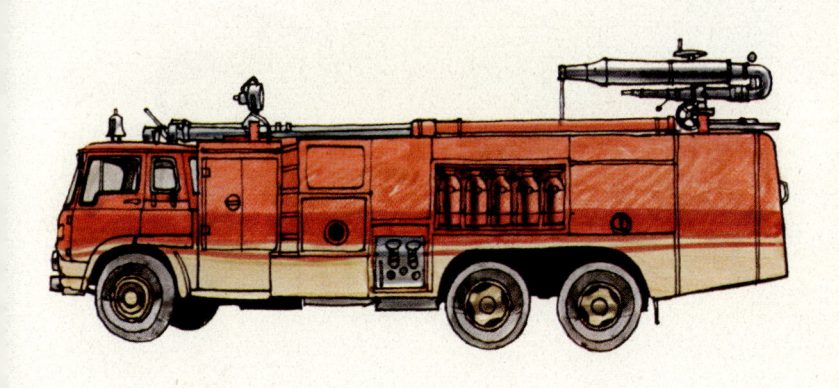

113 Bedford-Pyrene CO_2 and foam tender (UK)

114 Bedford-Pyrene 4×4 foam tender (UK)

115 High-output foam tender for oilfield use by Scammell-Pyrene (UK)

116 Pyrene trailer-mounted monitor (UK)

117 Bedford-Pyrene pump water tanker (UK)

118 Pyrene wheeled CO_2 extinguishers (UK)

119 Alvis Salamander 6×6 for the RAF (UK)

120 Carmichael Redwing, using forward-control version of Land-Rover (UK)

121 Land-Rover light fire appliance, the HCB-Angus Firefly (UK)

122 Maxim Mini-Max 2,800, 1971 (USA)

123 Maxim FF-CLT 100-ft aerial ladder, 1971 (USA)

124 Maxim–International 4×4 brush truck for forest fires (USA)

125 Leyland Firemaster pump, 1958 (UK)

126 Dennis F2 Pyrene foam tender, 1948 (UK)

127 Dennis F7 pump escape (UK)

128 Dennis F series Metz turntable ladder (UK)

129 AEC-Merryweather Marquis pump (UK)

130 Hino six-wheel turntable aerial ladder, 1971 (Japan)

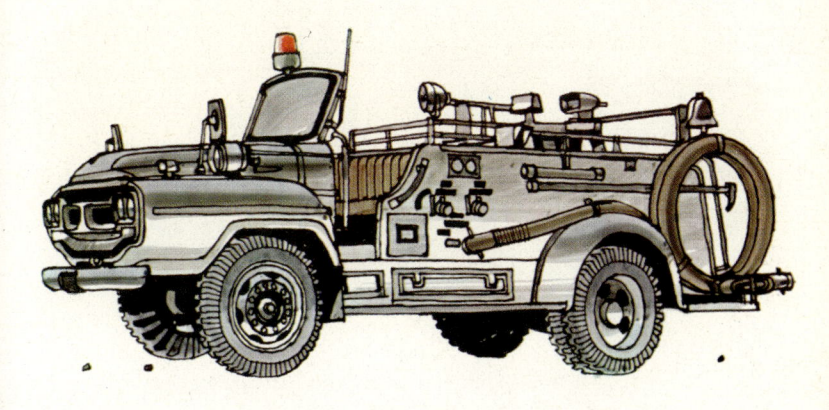

131 Isuzu light pumper, 1971 (Japan)

132 Toyota light pumper, 1971 (Japan)

133 Thornycroft Nubian II dual pumper, 1958 (UK)

134 Pyrene Protector on Thornycroft chassis (UK)

135 Thornycroft Nubian 6×6 airfield crash tender, 1960 (UK)

136 Willeme-SIDES LSP 6×6 airfield tender (France)

138 Mack tractor with monitor and tender (1965), part of Super Pumper Complex (USA)

137 Mack tractor with Napier Deltic engine and De Laval pump housed in trailer, 1965 (USA)

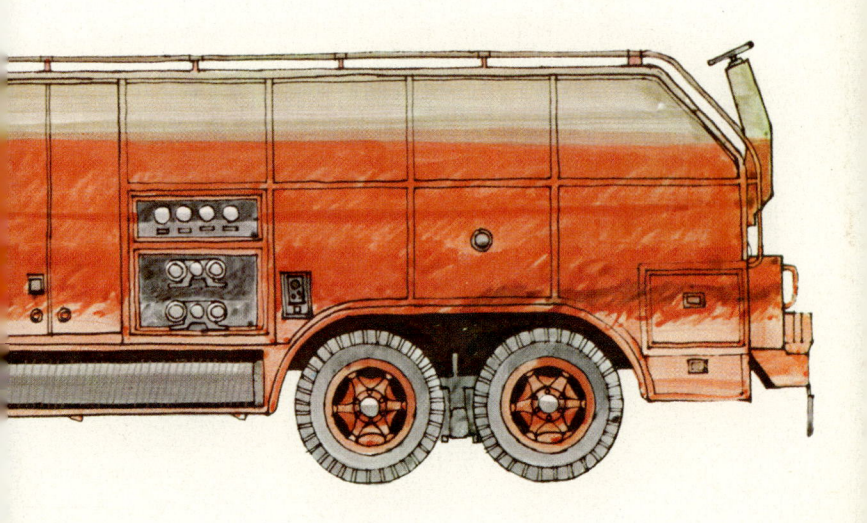

139 Mack Satellite tender of Super Pumper Complex, 1965 (USA)

140 Berliet GBK 18 4×4 forest fire fighter, 1971 (France)

141 Prototype fully amphibious fire pump by Eisenwerke, 1968 (Germany)

142 Skoda ASC 16 pump, 1962 (Czechoslovakia)

143 Scania-Vabis LB 7650 hydraulic platform, 1966 (Sweden)

144 Albion-Carmichael Firechief pump escape (UK)

145 Bedford Type B water tender (UK)

146 Zuk A14 light-type pump, 1968 (Poland)

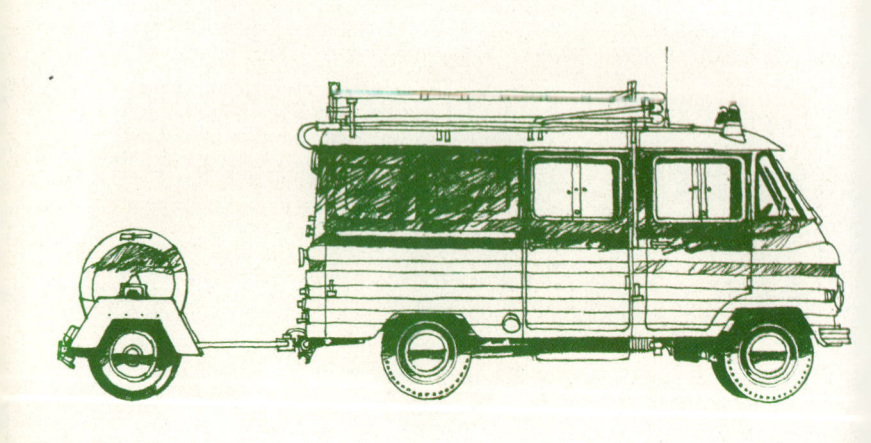

147 Zuk A15 light-type with trailer hose reel, 1968 (Poland)

148 Mowag-Chrysler airfield tender, 1972 (Switzerland)

149 Mowag 4×4 cross-country tender, 1972 (Switzerland)

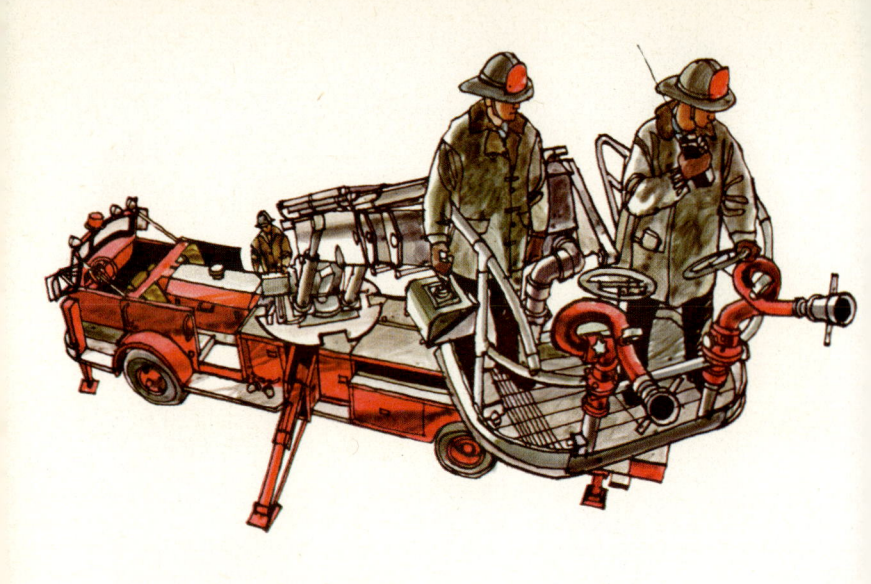

150 Mack Aerialscope, viewed from above (USA)

151 Mack Aerialscope (1972) in closed position (USA)

152 АЦ-20 4×4 pump water tender, 1972 (USSR)

153 АЦ-40 6×6 pump water tender, 1972 (USSR)

154 Ward La France Command Tower, 1972 (USA)

155 АЛ-30 6×6 30-m. aerial ladder, 1972 (USSR)

156 Faun LF 1410 8×8 1,000-h.p. foam airfield crash tender (1972) for the jumbo-jet age (Germany) (above: dry powder version)

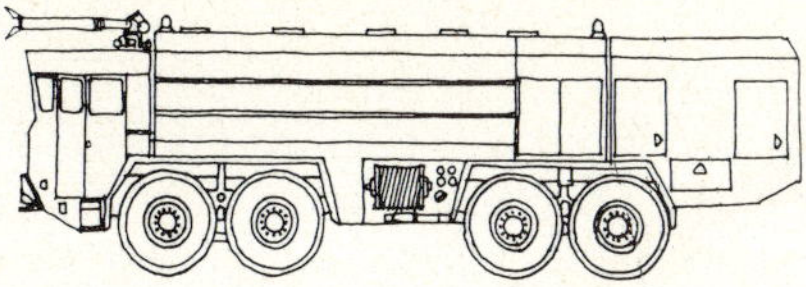

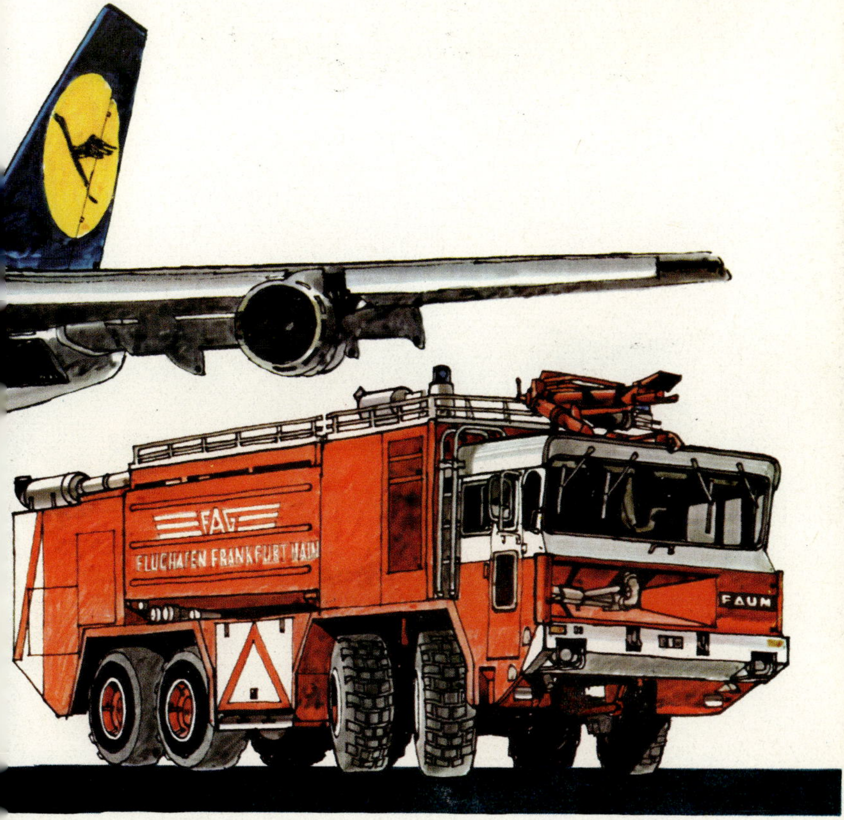
FAG
FAUN

157 Reynolds Boughton-Pyrene Pathfinder, 1972 (UK)

158 FWD-Kronenburg 6×6 airfield tender, 1972 (USA/Holland)

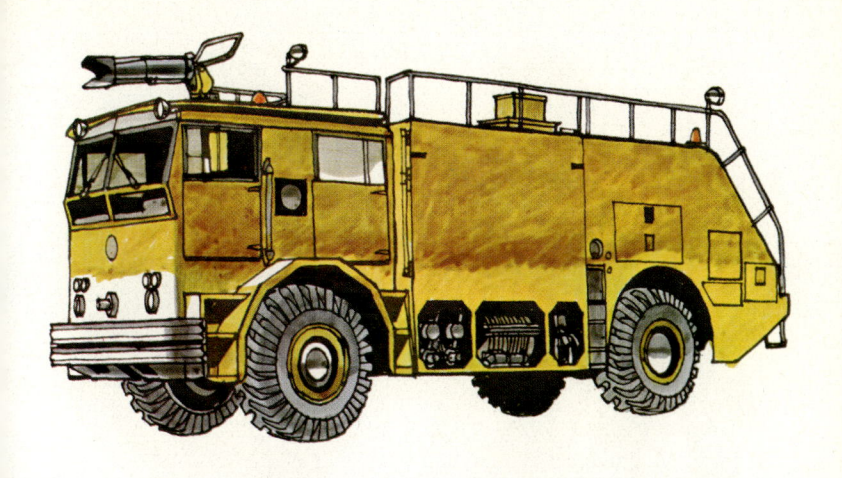

159 Ward La France 4×4 airfield crash tender, 1971 (USA)

160 Kaelble-Kronenburg 4×4 airfield tender, 1971 (Germany/Holland)

161 The modern fire fighter, British, Polish, American and British

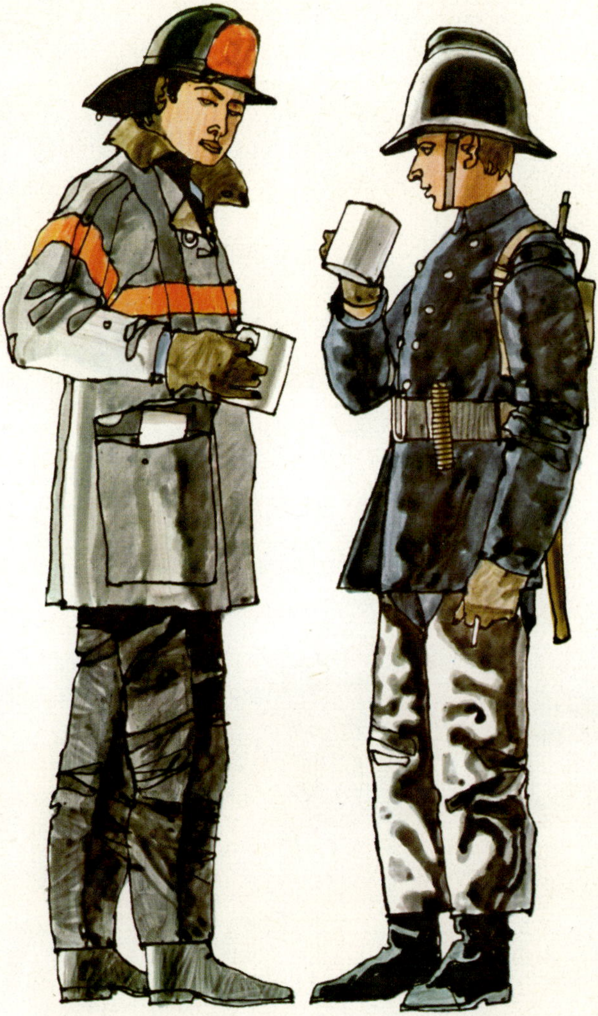

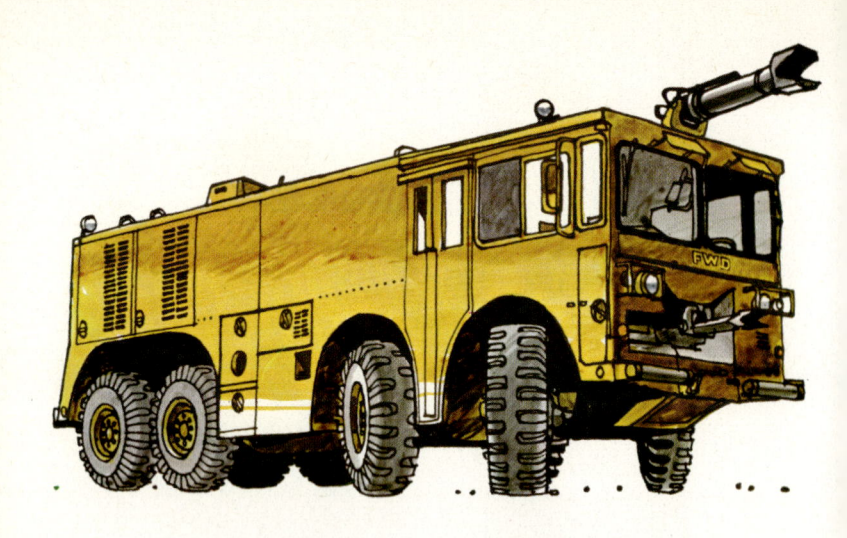

162 FWD P2 8×8 airfield tender, 1966 (USA)

163 Ford-Pyrene B-type water tender, 1971 (UK)

164 Steyr-Rosenbauer TLF2000, 1971 (Austria)

165 Csepel-Ikarus 344 pump water tender 1968, (Hungary)

166 American La France Pioneer II, 1972 (USA)

167 Range Rover-Carmichael Commando 6×4, 1973 (UK)

DESCRIPTIONS

Early Fire Fighters *Plate 1*

In fig. C we see a small band of early fire fighters getting to work with the squirt type of hand pump depicted in plate 1.

This basic principle or design was known long ago and was referred to as a syringe in the writings of Hero of Alexandria. The Romans used a similar instrument and called it a siphos and ordered all householders to have one for fire extinction. They were in general use until the seventeenth century, being located in churches, colleges and workshops. Cast in brass they consist of two parts, an outer barrel and an internal plunger. The barrel of the squirt had two large hand holds cast into its shape and these were grasped by the two men bearing the weight of the appliance. A third man took hold of the handle at the end of the inner plunger and used all his strength to draw back the plunger when the end of the squirt was immersed in the water supply. As the squirt was then aimed at the fire by the two men holding the barrel the man operating the apparatus pushed the plunger home with all his weight.

It must have been a tedious business fighting a fire in this manner because of the constant refilling of the apparatus and the need to rely on others to replenish the supply of water to the cistern being used for filling the squirt.

Fig. C Early fire-fighters with a hand squirt (UK)

Lucar's Manual Engine
Plate 2

It is possible that attempts were made to update and improve the squirt type of fire apparatus, and no doubt many variations were tried. To increase the output of such appliances without incurring the penalty of great weight, which limited the flexibility of manual machines, was of paramount importance. If a small body of men was going to be able to use the appliance it must remain fairly manageable with regard to size.

Therefore the design of a mechanised squirt, if one can call it that, attributed to Lucar during the middle of the sixteenth century, is worthy of study.

As will be seen from the illustration (plate 2) the apparatus consisted of a large cone-shaped cylinder with fixed spout and top filling funnel. At the large open end of the cylinder is positioned the plunger or piston which is a watertight fit inside the cylinder. The plunger is driven forward into the cylinder by the action of a coarse threaded screw turned by two men.

The principle of the action is that the cylinder is filled by buckets through the filling funnel with the piston at the back of the cylinder. When the cylinder is full of water the cock at the base of the filling funnel is closed, the men turn the screw as quickly as possible, thereby ejecting the water in the direction of the fire.

From the early drawing available of this apparatus it is not clear what shape the piston took nor the length of stroke possible. In fact the whole design is a little hazy although some detail may have been lost during the translation carried out in the sixteenth century. Nevertheless it is interesting to note what was being done during this period to increase the effectiveness of primitive equipment.

The Fire of London
Plate 3

We make no excuse for including just one illustration of a fire in this book and for sheer fury the Great Fire of London of 1666 could hardly be bettered. It must have been a dreadful happening for Londoners of the day. When we consider the puny methods of fire extinction existing at that time there is no doubt that London was extremely lucky to have survived at all. This four-day fire destroyed about two-thirds of the city and the history books quote the devastation as including 400 streets, 89 churches and 13,000 houses.

In the history of fire fighting this great fire marked an important point because so far as Britain was concerned much more attention was focused on fire protection, fire fighting and fire insurance than ever before.

In the year following the Great Fire a physician and builder named

Dr Nicholas Barbon engaged himself with rebuilding operations in the gutted City but, what is more to the point, he also set about promoting a scheme for fire insurance of houses and other buildings. The doctor was joined by others a few years later and the venture became known as The Fire Office. The Office adopted the badge of a Phoenix rising from the flames and so began the fascinating story of fire marks referred to in plates 4 to 15.

Our illustration (plate 3) shows London burning at the height of the blaze in a view from the south bank of the Thames.

Insurance Fire Marks *Plates 4–15*

To see a building today with its fire mark intact is something of a rarity. One must go to the offices of insurance companies, or to museums, or find private collections in the hands of manufacturers or individuals to see the wide and colourful variety of fire marks.

Under the original concept of fire insurance those premises which were insured against fire risks with a particular fire office displayed the appropriate fire mark. This fire mark had a dual function. It informed all and sundry that the premises were under cover of insurance by that office. It also advised the fire fighters as to which buildings were covered by their particular office.

As already mentioned, the very first fire mark or badge was the Phoenix rising from the flames adopted by The Fire Office about 1680. From 1705 this office became known as the Phenix.

The early marks placed on buildings were made of lead while later copper sheet was used. Later still iron and tin marks were used by some companies, while on a few occasions, where the premises were insured right from the time of building, the mark was moulded in terra cotta or carved in stone. For almost 200 years these colourful marks could be seen adorning the walls of various buildings, some premises having several marks where the risk was shared by various offices or where the contents were insured separately.

In many instances the fire mark also bore the number of the policy under which the premises were insured and these numbers were punched out of, or stamped or painted on, the mark plate.

Some of the fire offices discontinued the issue of marks around the middle of the nineteenth century while others continued to issue them much later, although in many instances they were no more than advertising signs.

The use of fire marks was not peculiar to England. Fire marks were used in many countries abroad from Portugal to China.

It will be noticed from the display of fire marks arranged in plates 4–15 that not all marks carried the name of the issuing company.

Those illustrated are:

Norwich Union Fire Insurance Society Limited	Established 1797
Kent Fire Insurance Company	1802–1901
Hand-In-Hand Fire and Life Insurance Society	1696–1905
Westminster Fire Office	1717–1906
Birmingham District Fire Office	1834–1864
Queen Insurance Company	1857–1891
Bristol Crown Fire Office	1718–1837
Protector Fire Insurance Company	1825–1836
Salop Fire Office	1780–1890
Lion Fire Insurance Company	1897–1902
Sun Insurance Office	Established 1710
Bristol Fire Office	1769–1840

Early Firemen *Plate 16*

As mentioned earlier the first fire insurance company was The Fire Office established in 1667. This office employed a small number of Thames watermen as fire fighters and so formed the very first insurance fire brigade. They were fitted out with distinctive livery uniforms and provided with arm badges showing the company to which they belonged.

It is significant that watermen were chosen as being the type of man required to carry out the duties of firemen, being men respected for their skill, courage and robustness when dealing with barges and their cargo. Many years later the tradition of employing ex-seamen was still prevalent in fire brigades.

The five firemen shown in plate 16 represent those of the many insurance brigades which existed during the 200 or so years from 1667. The fireman on the left is in the livery of the Royal Exchange brigade, which was formed in 1720 and joined the Sun and Union in 1825 to form a joint brigade. Next to him is a member of the London Assurance brigade, also founded in 1720, and next to him is a fireman of the Atlas brigade which, together with ten other insurance brigades operating in London, were joined together in 1833 to form the London Fire Engine Establishment. The fireman standing armed with a preventer is a man of the Sun Fire Office brigade which was formed in 1710, while last in the group is a fireman of the Westminster brigade which was formed in 1717 and joined with the Alliance in 1906.

Manual Engines by Newsham and Others *Plates 17–19*

One of the oldest fire pumps existing today is that shown in plate 19. This tub-shaped pump was originally supplied to the Bedfordshire town of Dunstable around 1678 and is believed to have been built by Keeling of Blackfriars. The pump is constructed as an oval wooden tub some 4 ft long by 3 ft wide, placed on a wooden frame supported on four wheels of 2 ft 5 in. diameter. Overall dimensions are length 10 ft, width 3 ft 6 in. and height 6 ft 6 in. The pump is now preserved in the London Museum.

The second tub-type machine illustrated in plate 17 is somewhat similar in design to the Dunstable appliance but in fact originates from Belgium, where it was still at work in 1914. In this machine the handles are made of metal and a swivel-jointed metal nozzle is provided for projecting the water in the required direction.

In Britain it has been generally acknowledged that the design of fire engine patented by Richard Newsham in 1721 marked a great step forward in bringing the manual engine to perfection. To say that Newsham invented the manual fire engine is absurd. He merely brought the principle up to date by introducing features in the design and construction which gave an improved performance. No doubt Newsham had learned something from the engines produced by the brothers Van der Heyden in Holland, but irrespective of this, the Newsham design gave the manual engine the new lease of life it badly needed.

We must remember that even in this era the engines were handpropelled with no means of steering. For greater flexibility the pump had to be mounted in a horse-drawn cart. It was left to Hadley and Simpkin to produce an engine with road springs and a locking carriage later in the nineteenth century, which enabled the pump to be more easily handled.

The machine illustrated in plate 18 is one of the Newsham engines built in the 1730s, having two single-acting pump barrels with an air vessel to maintain an even discharge. The bottom ends of the pump barrels are within the tank forming the base of the machine, and a two-way cock is provided so that water can be drawn into the tank by way of a suction hose or the tank be replenished with buckets.

One idea that Newsham had about fire engines was that one should arrange the pumping facilities in such a way that as many men as possible can be set to work on the pumping operation. This he did by placing the pump handles along the sides of the machine rather than at the ends, as other builders had done. He also provided treadles in the centre of the machine so that additional men could throw their weight into the pumping rhythm by working with their legs. These pumpers were assisted in their task by the long fixed rails on top of the machine, which gave it the appearance of a four poster bed.

Fig. D Newsham manual pump at work (UK)

In fig. D we see a Newsham engine at work with pumpers at the side handles and on the treadles, while another crew member holds the swivelling metal play pipe and a willing lady tops up the water in the tank with a bucket.

Tools, Equipment and Ladders *Plates 20-2*
In the various descriptions of the fire engines in this book there will frequently be found a short list of the various small items carried by appliances. In many instances these are too numerous to mention individually and merely consist of the tools and items of equipment which transform the bare carrying vehicle into a copiously equipped fire-fighting and rescue machine. Without them the crew would be hard pressed to carry out their job efficiently and expeditiously. Often the problem is not how many items should be carried on the appliance but whether the next emergency will demand some tool or device which is yet to be perfected.

Let us mention some of the equipment which has been carried on appliances or used on its own.

The early small item was of course the bucket and these have been produced in canvas, leather, wood or metal. The collapsible canvas ones with rope handles are useful for storing in a small place.

Portable dams or cisterns also made of canvas were of great use in keeping small manual appliances going, and these have been produced in wood or metal as well.

A preventer is an item of equipment that has been in use for many years, having its origins in Roman times. It is useful for making holes in ceilings, pulling tiles off roofs and many other tasks beyond the reach of one's hands. (Plate 20c.)

The axe or hatchet is another ancient item of equipment useful for gaining quick entry through locked doors, windows, roofs, pavement lights, etc.

Crow bars perform a similar function and find use in forcing open doors, windows, roof lights, gates, etc. Doors of crashed vehicles often have to be wrenched open, railings have to be parted to release trapped children and hatch covers forced off to gain access to ship's holds.

On many appliances saws of various types have been carried – rip saws, hacksaws and, of recent introduction, power saws for gaining quick entry to wrecked cars.

Lamps are a most important item to the fireman whether it be on the appliance going to the fire or for his own personal use when searching for missing persons in a building. Many appliances have carried a swivelling spot light, others have detachable lights with cable reels and stands for use at some distance from the appliance. So important is light to the fireman that special appliances have been produced solely for providing adequate illumination at fires. Earlier appliances used independent pressure gas burners while later models use a special generator and extension leads. An early type of lamp is the tubular naphtha torch shown in plate 20b. This not only provided light for the fireman but was used to light up the fire of the steam pump.

The jumping sheet did not find great favour in Great Britain, but in America and the Far East has been used on many appliances. In fig. G a folded jumping sheet is shown stored on board an American appliance, while on some other appliances the sheet was made of strong canvas with a rope edging for easier storage.

Ropes and lines are another adjunct of the fireman's equipment and have a variety of uses including lifesaving and rescue.

Hydrant fittings, standpipes and keys are carried on machines which may have to use water from street mains.

Nozzles, branches, breechings and hose couplings are items associated with the delivery hose. Nozzles are the small-diameter metal ends of branch pipes and are the last controlling factor in determining the jet. A branch pipe is the tapered metal tube which is held by the fireman on the end of the delivery hose. Some time ago various designs for controlling the shape of the water jet were produced. Plate 20a shows a fan spreader for producing a thin fan of water over a large area. Breechings are the metal sections employed in dividing the streams of water in a hose. Hose couplings as the name implies are the metal connectors used to connect successive lengths of hose.

Next to hose and water the ladder is probably the most widely used item on a fire engine. Ladders have appeared in a wide variety of types, materials and sizes, ranging from short hook ladders, through two- or three-section extension ladders, to extending escapes and long aerial ladders. Materials used have included wood, steel, aluminium, stainless steel, combinations of wood and metal and mixtures of varieties of metal. Ladders have been extendable by hand, rope, windlass, electric power, compressed gas, hydraulics or mechanical means.

To illustrate the variations on a single theme a selection of hook ladder designs are shown in plate 21, while in plate 22 is shown a particular type of detachable wooden escape – the Wivell fly ladder escape. This type of rescue escape consisted of a main ladder 31 ft long mounted on a two-wheel carriage for manoeuvrability with a pair of small wheels on rollers at the head of the ladder to assist in positioning against a building. Near the top of the main ladder was hinged a shorter ladder 19 ft long which was positioned by ropes. Beneath the main ladder was a wire mesh shoot down which rescued persons could slide to the ground.

Steam Fire Engines *Plates 23–30*
The first steam fire engine in Europe was constructed by Mr John Braithwaite of New Road, London, in 1829 (plate 24). He was assisted by a Mr Ericsson and together they produced a two-cylinder engine producing 10 h.p.

The general layout of the engine was a boiler stoked from the rear of the machine which fed the two horizontal cylinders. The exhaust steam was led from the cylinders through the feed water tank as an aid to preheating the water. The two pistons of the engine were connected directly to the two plungers of the two-cylinder pump which was at the front end of the vehicle just behind the driver's seat. A large spherical air vessel surmounted the pump, and connections for the suction and discharge hoses were placed just beneath it.

The complete machine weighed only 45 cwt and pumped at the rate of approximately 150 gals per min. Coke was used as the boiler fuel.

Some two years later a second steam fire engine was produced at the same works, to a similar design as the first engine but of smaller output, being only a single-cylinder machine of 5 h.p.

The third Braithwaite engine is shown in plate 23 and follows the same general layout of his first machine in having a two-cylinder horizontal engine, but in this case gearing was introduced between the pistons and the pumps. The machine also differed in that it had three pump barrels instead of two as in the first appliance. This engine went to work at Liverpool.

Braithwaite's fourth engine was the *Comet* built for the King of Prussia in 1832 (plate 25). This 15-h.p. machine was of the two-cylinder, two-pump type and weighed 4 tons. Output was reported as being approximately 300 gals per min.

The fifth and last Braithwaite steam fire engine was built in 1833.

It was Mr Paul R. Hodge who built the first steam fire engine in America. It was constructed in his New York works during the period December 1840 to April 1841 and was based on railway locomotive practice. It will be noticed in plate 26 that the large-diameter rear wheels are jacked up off the ground. It is in this position that the engine is put to work, the road wheels acting as flywheels. The boiler provided steam for the two horizontal cylinders which were placed on either side, the piston rods driving through the cylinders to the double-acting pumps bolted directly on to the cylinders.

It is reported that in 1851 W. L. Lay designed a self-propelled steam fire engine and that in 1853 a 'self-propeller' steam fire engine was built by A. B. Latta. This vehicle was a three-wheeler. 1855 saw a machine constructed by Abel Shawk of Cincinnati, with a water tube boiler feeding a single-cylinder engine which powered the single-cylinder pump.

In 1858 Messrs Poole and Hunt of Baltimore started building steam fire engines and plate 27 shows one of theirs constructed in 1865. This machine had a vertical boiler and single-cylinder engine but a two-cylinder pump arranged with the cylinders one above the other.

Returning to Europe we find that in 1858 Messrs Shand Mason & Company built their first steam fire engine (although they had gained experience earlier in the building of two steam fire pumps for floating engines of the London Fire Engine Establishment).

In 1860 James Sheklton built a steam fire engine in Ireland, and in 1861 Merryweather & Sons produced their first, *Deluge* (plate 28). This engine was of vertical boiler design with single horizontal cylinder and double-acting pump, and took part in the very first trial of steam fire engines held in Hyde Park, London, during the 1862 International Exhibition. It is reported that the machine cost £700 and weighed 65 cwt.

A year later, in July 1863, the trials took place at the Crystal Palace and again Merryweather were present with two engines, *Torrent* and *Sutherland*. The *Torrent* was a single cylinder and pump type and was entered in the small engine class (plate 29). The *Sutherland* was a double horizontal cylinder and pump design and was placed in the large engine class in competition with six other engines.

The first self-propelled steam fire engine in Europe was the three-wheeled machine built by William Roberts of Millwall, London, in 1862. Measuring 12 ft 6 in. long and 6 ft 4 in. wide it weighed $7\frac{1}{2}$ tons and achieved an average of 14 m.p.h. The engine was a two-cylinder vertical 6 in. × 12 in. and the pumps were two double-acting pumps 9 in. × 8 in. The two driving wheels were 5 ft in diameter while the single steered front wheel was 3 ft in diameter. (Plate 30.)

Later in 1862 the second Roberts engine was built – *Princess of Wales* – and this machine took part in the 1863 trials at Crystal Palace.

Pirsch Hook and Ladder   *Plate 31*

The name of Pirsch ranks among the pioneers in the field of fire-fighting apparatus builders in the United States. Although the company is best known for its metal ladders of various types, it has in fact built apparatus of all kinds and is still in the leaders' class of equipment today, building pumpers and aerial ladders on both rigid and articulated chassis, front or rear mounted, as well as equipment on proprietary chassis including Snorkels.

The first patent relating to fire equipment was taken out in 1899 when Peter Pirsch of the Nicholas Pirsch Wagon and Carriage plant was granted a patent for his design of compound-trussed extension ladder.

In 1900 the Peter Pirsch Company was founded and hand- and horse-drawn ladder trucks were built. In 1910 the first Pirsch-equipped chemical and hose car type of fire apparatus was delivered to Kenosha, while in the following year the first motorized piece of equipment was produced – a city service ladder truck. Things continued to improve for the company and by 1916 the first Pirsch-equipped motorized pumper was produced for Creston, Iowa. Later in the same year a motor pumper was delivered to the Chicago Fire Department.

There followed a variety of motor-driven fire apparatus including ladder trucks, pumpers, aerial ladders and hose cars as well as certain special vehicles such as smoke ejectors. One of these pieces of equipment was instrumental in saving the lives of sixteen Chicago firemen in a 1931 fire.

1926 saw the introduction of the first Pirsch-built motor chassis, this carrying a 500-gal. pumper body and equipment. Later in the same year a 600-gal. pumper was built, and this was followed by a 750-gal. machine in 1927. The first 1,000-gal. pumper built was delivered to the City of Fond du Lac, Wisconsin, in 1928 and not long afterwards apparatus embodying 1,250- and 1,500-gal. per min. pumps were offered.

The Pirsch company hit the headlines in 1931 when the world's first all-powered aerial ladder truck was produced and sold to the city of Spokane, Washington. This triumph was followed by another in 1935 when Pirsch produced a 100-ft aluminium alloy aerial ladder. Not only was this particular ladder the first all-powered ladder with an extension in excess of 85 ft in America, but it was also the first all-powered metal aerial ladder of any length. This machine went to the city of Melrose, Massachusetts.

Plate 31 shows a Pirsch hook and ladder wagon of the late nineteenth century, and a modern aerial ladder is shown in fig. E.

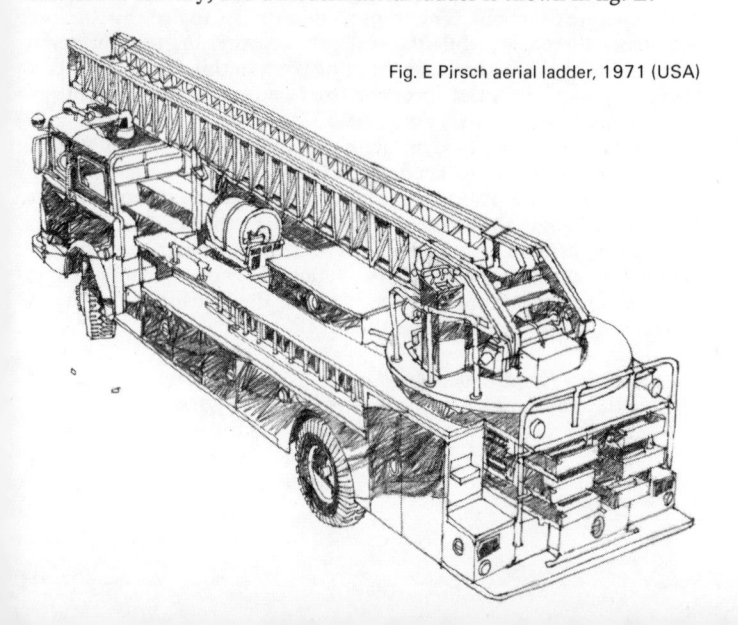

Fig. E Pirsch aerial ladder, 1971 (USA)

Horse-Drawn Equipment *Plates 32–4*

As mentioned earlier the first horse-drawn fire engines were only ordinary horse carts with the old manual engine carried as a load. It is not until about 1800 that we find the engine itself being fitted with shafts and locking gear so that it could be moved as an independent vehicle.

The earliest horse-drawn fire engines were the manually pumped type with pump handles mounted at both sides. When the early steam pumps appeared they were fitted with either shafts for one horse or a pole for two or more horses.

Other types of vehicle were introduced for use with horse traction and hose carts, ladder trucks, escape carriers, turntable ladders, water towers, canteen vans, water tanks, chemical apparatus and many combination wagons appeared.

An American horse-drawn steamer is shown in plate 32. It is interesting to note that many of the steam pumps produced toward the end of the horse-drawn era were regarded as indispensable for fire fighting well into the motor vehicle age. In order to retain them some brigades dispensed with the horse pole and shafts and fitted a towing bar so that the pump could be moved by a motor vehicle.

In plate 31 is shown another American wagon in the shape of a hook and ladder truck produced by Pirsch toward the end of the nineteenth century.

Of old design but late construction is the duplex machine produced in 1906 by Sternberg 'es Kalman of Bucharest. This interesting type is almost two separate two-wheel carts joined together by means of the central spine or coupling pole, and the two ends of the vehicle have separate functions. The front section carries the crew with hose reel under their seat, while the rear part supports the manual pump and lengths of hose. Another interesting point in this design is the fact that the pump is worked from the front and rear and not from the sides, as most late manuals were (Plate 33).

Another interesting type of late horse-drawn engine is that shown in plate 34. Produced by Teudloff Dittrich of Hungary, this machine really does show the transition between horse power and motor power because although the wagon is propelled by horses the pump is powered by a small petrol engine! With 200-gal. water tank, motorized pump, extension ladder, hard suction hose and a small hose reel this type of vehicle must have been one of the last of its kind when produced toward the end of the 1920s.

Early London Fire Brigade Appliance *Plate 35*

During the closing years of the nineteenth century several attempts to produce self-propelled vehicles were made and the goal was always to produce a vehicle which was powerful, strong and reliable but without incurring the penalty of great weight. The Roberts self-propelled steamer of the 1860s had provided a lead of a sort, but no one was able to follow this with an improved design until the 1890s, when Merryweather produced their self-propelled steamer for Port Louis.

About the same time some electric fire engines came onto the scene in Britain, France, Germany and the United States. A few chief officers were trying out early motor cars (although these are not strictly fire engines) as a means of getting to the fire speedily, and it is reported that in 1900 Chief Croker of the New York Brigade was accused of being a public menace as he raced to a fire at 60 m.p.h. in his Locomobile!

In London the Metropolitan Fire Brigade took a 300-gallon horse-drawn steamer and converted it to self propulsion in 1902, but it did not prove a success because of insufficient boiler capacity. In 1906 the machine was changed back to horse traction.

In 1903 another idea was tried. This was to obtain a four-cylinder 24/30 h.p. Wolseley tractor unit and couple it to a horse-drawn steam pump which had the forecarriage removed and the front end superimposed on the rear of the Wolseley. The resulting articulated outfit was far from a success. Most of the weight of the steam boiler and pump was above the iron-tyred wheels of the trailer portion and this, together with the fact that the trailer wheelbase was shorter than that of the tractor, made for a most unmanageable vehicle.

Also in 1903 a two-cylinder 10/12 h.p. Wolseley on pneumatic tyres was purchased for conversion to a hose tender by the brigade workshops. The resulting vehicle (plate 35) had seats for a crew of three, a small pump driven off the engine and a hose box for 500 ft of delivery hose. The pump provided insufficient output when tried and was removed but the vehicle remained in service as a hose tender for several years.

Merryweather Fire King   *Plate 37*

In 1690 Nathaniel Hadley established his fire engine business in London and after a short while was joined by a Mr Simpkin. Many years later, in 1791, Henry Lott, son of Squire Lott, joined the partners to establish the business of Hadley, Simpkin and Lott. In the following year patent No. 1925 was granted to the firm in respect of a fire pump.

The name Merryweather first appears on the scene in 1807 when young Moses Merryweather joined as an apprentice. He remained with the company, married Lott's niece and in due time took over the company when Lott retired. The company later became Merryweather & Sons.

With a long history of producing manual and steam pumps Merryweather produced their first self-propelled steamer in 1899. This machine was fitted with the Gem fire pump and marked the start of about ten years of steam vehicle building.

The type became known as the Fire King and was available in a series of six basic models with various options relating to fuel and wheel types.

The machine consisted of a straight parallel frame mounted high off the ground. The front panel was steel, and axles wrought iron and the hose box and seat, mahogany. The patent Merryweather vertical water-tube boiler was positioned just in front of the rear axle with two 50-gallon water tanks positioned on either side.

The double-cylinder vertical engine was mounted just in front of the boiler and could be connected to rear wheels or pump as required. The Merryweather Gem two-cylinder pump was mounted in front of the boiler, directly beneath the engine. If an oil-fired boiler was fitted fuel was stored in tanks carried beneath the hose box.

Transmission was by means of a countershaft driven by cog wheels engaged by a positive clutch, carrying sprocket wheels and having differential gear. Drive was then by steel roller chains to sprockets on rear wheels. Wood wheels with iron tyres, wood wheels with solid rubber tyres and wrought iron wheels with steel tyres were offered as alternatives.

Prices varied from £850 for a 300-gallon machine with coal-burning furnace, wood wheels and iron tyres, up to £2,275 for a 1,000-gallon machine equipped with oil-burning furnace, wrought iron wheels and steel tyres.

Electric Turntable Ladder   *Plate 38*

Although the first turntable ladders were horse-drawn it was not long after the advent of the motor age in fire vehicles that the attention of engineers was drawn toward producing self-propelled sets of ladders.

Because of their weight and high centre of gravity a low-built machine of ample power and reliability was required for turntable ladders. With the early motorized chassis they were rather heavy and high off the ground and not always reliable. So it was the electric vehicle that came into use for ladders during the early part of the century.

Soon after 1900 the Vienna-based manufacturer Lohner-Porsche started to produce electric and petrol-electric vehicles, and a few fire engines were built during these early years. The Lohner-Porsche design included electric motors built into the front wheels of the vehicle and this feature was carried on when a few years later the Mercedes Electrique appeared. Mercedes having taken over the Lohner-Porsche production marketed the design in Great Britain as the Gearless Electric.

Another supplier of electric fire engines during these early years was Henry Simonis who after acting as chief representative of the German fire engine manufacturers at the 1903 Fire Exhibition at Earls Court set up business in London. In 1906 Henry Simonis supplied an 85ft turntable ladder to Glasgow which had a Mercedes 30/40-h.p. engine producing electric current to two front-wheel-mounted electric motors. Power for the ladder was provided by compressed carbon dioxide gas. This machine weighed $6\frac{1}{2}$ tons in working order and could travel at 20 m.p.h. Unlike the Lohner-Porsche design this chassis had the electric motors driving the front wheels by means of an internal gear ring and not incorporated into the wheels themselves.

The electric turntable ladder illustrated in plate 38 is a type marketed by Cedes Electric Traction Limited on a chassis produced by the Austro-Daimler concern. It incorporates a ladder taken from an old horse-drawn appliance. This design incorporates the electric motors built into the front wheels, the power for which was supplied by batteries stored in a box under the driver's seat. These early turntable ladders were principally rescue appliances and carried no pumps or hoses. The short ladder carried at the side of the main ladder is used to reach the bottom section of the ladder when it is in the elevated position.

German Steam Fire Engine *Plate 39*

It was a natural progression from the steam fire pump to the self-propelled steamer, but for a variety of reasons not as many steam-propelled engines were produced as one would have expected. The first European self-propelled steamer was built by William Roberts in Millwall in 1862 (plate 30). Contrary to expectations there was no rush of competition nor flood of orders. Fire engines were expensive machines, the steam pump more so, and as for buying a self-moving engine which would probably stand idle for long periods when there were always plenty of good horses to pull it, that was out of the question.

It was not until almost the end of the nineteenth century that a few self-propelled steamers did begin to appear, and then it was only a handful. By this time it was almost too late for the young steamer to reach maturity because the infant petrol engine was gaining ground in light vehicles at an alarming rate, while the brother electric vehicle was also being tried to an ever-increasing degree.

Nevertheless some interesting steamers did appear, and in plate 39 we see one of the German machines which took to the road in 1905. This was turned out from the works of Waggon und Maschinenfabrik at Bautzen. The vehicle has a very high frame so that the forecarriage turns beneath it. The large-diameter rear wheels carry most of the weight, for the vertical boiler, engine, pump, coal box, water and oil tanks are situated at the rear of the vehicle. The crew are carried on a bench seat at the front, which doubles as a tool box, and on top of the large water tank just behind it. The engineer or stoker stands at the rear between the coal box and the small water tank.

At the time this machine was delivered it was reported that the builders were still not financially successful. This serves to illustrate the fact that there was a limited demand for such appliances even in 1905, and in a few years the position was to deteriorate even further with the rapid progress in the design of the petrol-engined vehicle even for quite heavy loads.

Early German Fire Engines *Plates 40–4*

In the early part of the century the German vehicle builder Justus Christian Braun of Nuremburg built some most interesting fire engines on a variety of chassis. The goods vehicles produced were electric, petrol and steam types while the fire appliances seem to have appeared in various guises mounted on a battery-electric chassis.

The appliance shown in plate 40 is described in literature of the day as an 'elektromobil-steam' fire engine and features a high-chassis frame running parallel to a point just in front of the rear axle where it is curved outwards to accommodate the boiler for the pump. The two-cylinder reciprocating pump produces 1,500 litres per minute. It is mounted just in front of the boiler and draws its water from a 200-litre cylindrical tank which has a hydrant coupling for replenishment. The boiler uses paraffin fuel and carbon acid cylinders provide gas to assist the flow of fuel. Power for the vehicle is supplied by batteries stored in a large box which doubles as a seat for the driver. Drive is by two electric motors built into the front wheels. The two front wheels, motors, axle and road springs all pivot on the turntable which is turned by a horizontal steering wheel.

About 1910 Braun supplied two vehicles to the fire brigade at Charlottenburg and although these both used the basic front wheel electric drive chassis system, that was where the similarity ended. The appliance shown in plate 41 is of rather startling design in that the chassis consists of two large-diameter Mannesmann tubes which are utilized for carrying 500 litres of water! The crew of twelve were seated on four cross benches which were positioned so high that two long steps were required to gain access to them. The high floor level enables the battery box, a folding ladder and two hook ladders to be carried underneath. Cylinders of carbon acid are used to discharge the water carried in the vehicle and connections are provided for hoses and hydrants.

The two Siemens-Schuckert 7·5-h.p. motors could operate at up to 1,100 r.p.m. which gave a road speed of 28 k.p.h. With an all-up weight of 7 tons the driver must have had his hands full for the controls were hand controlled, hand brake and horizontal handwheel for steering!

The second vehicle supplied was the Braun-Schappler extending ladder shown in plate 42. Again the appliance was based on a battery-electric chassis embodying two-motor front-wheel drive with the traction batteries being stored in a large box mounted above the front axle turntable. The traction batteries were also used to turn and elevate the ladder while the power for ladder extension was provided by carbon acid stored in four pressure cylinders clamped to the ladder base frame. The ladder sections were not trussed but took their rigidity from the large-diameter tubes which formed a backbone to the ladder.

Another two types of early German appliance are shown in plates 43 and 44. With these two machines a common form of propulsion is used – batteries under bonnet driving electric motors housed in the front wheels. However, the pump on one machine is powered by a steam boiler at the rear while the appliance carrying the ladders has a petrol engine for pumping power.

Electric/Petrol Motor Fire Engine *Plate 45*
During 1911 two Braun electric/petrol fire engines were delivered to the Hanover Fire Brigade as part of that brigade's modernisation programme. The two machines were based on the Braun battery-electric chassis using a bonneted layout having front wheel drive. Two electric motors were used, one in each front wheel, and power was supplied by the batteries stored under the bonnet. At the rear of the vehicle was mounted a four-cylinder Argus petrol engine of 54 h.p. with rear-mounted radiator using fresh water from the pump circuit. This engine was for pumping duties only, being connected to the midships-mounted pump by means of a Leblanc-type clutch.

The Pittler fire pump was capable of delivering up to 1,000 litres per minute whilst turning over at 600 r.p.m. and had suction and delivery fittings at both sides of the vehicle.

Three of the crew of seven were carried on the bench seat, while the remainder sat on top of the hose locker in front of the pump compartment. Further hose was carried on the detachable hose drum mounted at the extreme rear of the machine, while the four lengths of hard suction hose were carried on the running board tool boxes.

Performance of these machines was reported to be about 20 m.p.h. with a range of approximately 30 miles on one battery charge.

One of the machines that was destined for Hanover brigade is shown in plate 45.

Laurin and Klement, 1913 *Plate 46*

In 1895 two young men, Vaclav Laurin, a mechanic, and Vaclav Klement, a publisher, opened a cycle repair shop in their home town of Jungbunzlau in the Czechoslovakian territory of the Austro-Hungarian empire. As they gained experience they turned to the manufacture of cycles and in 1898 to the building of motorcycles. 1905 saw their first automobile take the road and during the next twenty years cars, buses, lorries, road rollers, industrial engines and fire engines were produced by the partners. In 1925 Laurin and Klement was amalgamated with Skoda.

The fire appliance shown in plate 46 is a type produced in 1913. Based on the 4-ton chassis the vehicle was powered by a four-cylinder 6-litre engine which developed 32 h.p. Fire equipment consisted of a rear-mounted positive displacement pump with air chamber producing 1,400 litres per minute. Overhead gallows were provided for an extension ladder and just below this were hung the three lengths of suction hose. In addition to the two fixed hose reels rolled hose was carried in the box body on top of which sat six members of the crew on unusual angled seats.

Merryweather motor pumps *Plates 36, 47*

The first self-propelled motor pump in Britain was the Merryweather pump escape shown in plate 36. This was supplied to Finchley District Council Fire Brigade in 1904 soon after Merryweather had supplied a motor chemical appliance to another London fire brigade at Tottenham in 1903.

In its original form the appliance was equipped with a Hatfield three-barrel reciprocating pump of 250 gallon capacity, 60-gallon soda acid tank, 180-ft hose reel and 50-ft wheeled escape. Power was supplied by a four-cylinder 24/30 h.p. petrol engine with chain drive to the rear wheels which moved the vehicle at speeds up to 20 m.p.h. During its life as a fire engine the machine underwent several changes. The original engine was replaced with one of 50 h.p., the curved radiator was replaced with a flat one as used on later Merryweather engines, the steering mechanism was modified to give a raked column, and twin rear wheels were fitted for greater adhesion.

The machine remained with Finchley until 1928 when it was sold out of service. It saw service as a pump in a gravel pit for a short while until it was spotted by T. D. Barclay and acquired by the London Science Museum for exhibition.

Following on the supply of the Finchley machine Merryweathers made certain improvements to their motor engines using a heavier type of wheel and tyre, raising the escape gantry and fitting a powerful searchlight. A more powerful Aster 50-h.p. engine was fitted as standard on both motor pump and chemical engines. In 1908 a suitably modified chassis of this type was built to carry the first British turntable ladder using the vehicle engine to power the ladders.

1910 saw a new type of Merryweather appliance in the shape of a forward-control battery-electric first aid machine with chemical tanks, but the electric fire engine did not find favour in Britain and the idea died.

In plate 47 is a Merryweather motor pump of 1913 which featured the Aster 50-h.p. engine and a Hatfield 350/400-galls per min. reciprocating pump. It is interesting to note that this full-size appliance was originally delivered to a large brewery for its private brigade. It is now preserved by Michael Banfield.

Seagrave Pumper
The Seagrave company was formed in 1907 at Colombus, Ohio. It started by producing mainly petrol-engined fire apparatus, but one early design worthy of mention is an unusual front-wheel-drive electric which carried an aerial ladder the base of which was mounted over the front wheels in front of the driver!

A more orthodox Seagrave pumper is shown in fig. F, being a 1914 machine with midships-mounted pump. This vehicle was powered by

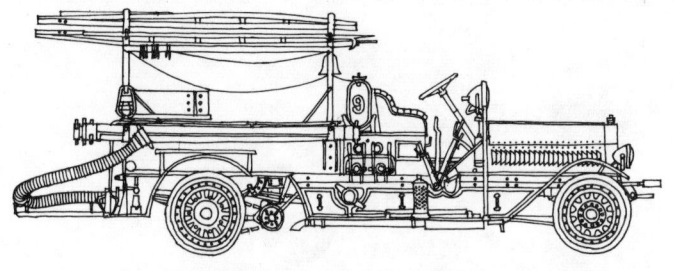

Fig. F Seagrave pumper, 1914 (USA)

a six-cylinder engine of $5\frac{3}{4}$ in. $\times$ $6\frac{1}{2}$ in. which produced 116 b.h.p., and drive was by way of a triple-plate clutch and three-speed gearbox to a countershaft and side chains to the rear wheels.

Although the almost standard type of American fire apparatus body is mounted on this machine it does have a somewhat British appearance because of the high ladder gantry. Most U.S. apparatus had a slat-bottomed open body for the storage of hose, and the ladders were hung on the sides in combination engines such as this.

American Motor Apparatus Plates 48, 53

The production of American fire apparatus today is mainly in the hands of a few large corporations together with a handful of smaller firms. As in so many fields, if we turn back the clock to the early days of motor apparatus in the first fifteen years of this century we find a whole host of names, many of which are still with us today in some form, while others are still well remembered and a few almost forgotten.

Some of the early motorised apparatuses were not fire engines at all, but were merely petrol-engined tractors built for coupling to old horse-drawn equipment which the brigade or company wished to keep up to date without having to resort to buying completely new equipment.

Under this heading came the Cross, the Klieber, the Christie, the Seagrave, the Mack and the American La France. In some cases the locking fore-carriage of the horse-drawn apparatus was removed and a four-wheeled tractor with rubbing plate and kingpin was attached. A Mack of this type coupled to an old horse-drawn water tower is shown in plate 53.

Another way of converting horse-drawn equipment was to remove the fore-carriage, driver's seat, footboard and controls and bolt a two-wheeled tractor direct to the frame of the apparatus. This gave a rigid four-wheel vehicle as opposed to the articulated type where a four-wheel tractor was used. In plate 48 an American La France 105 h.p. six-cylinder two-wheel tractor unit is seen coupled to an old steam pump. A less powerful four-cylinder 75 h.p. tractor was made by American La France for handling the lighter aerial ladders. In fig. G is one of these Type 31 tractors as attached to an old Hibernia hook and ladder apparatus. The wheelbase of this amazing machine is all of 31 ft 9 in., hence the need for a rear steersman. Note the folded jump sheet strapped to the side of the ladders.

Among the early motor appliances were such names as Locomobile, Pope-Hartford, Childs and Webb. Howe began in 1908, Waterous in 1907, and Robinson in 1911. Luitmeiler, Luverne, Universal and General were producing apparatus in the teens and Maxim, Moreland, Kenosha and Knox all followed around the same time.

Scania-Vabis Appliance   *Plate 49*
The Scania-Vabis enterprise in transport goes back to 1891 when the Vabis part of the present company was formed as a railway repair shop. The first motor vehicle was produced in 1897, the first truck in 1903 and the first bus in 1911.

The 1917 fire appliance shown in plate 49 was based on the 2-ton truck chassis of the period with a four-cylinder engine producing 30 h.p. from 2·81 litres. It was fitted with a rear-mounted pump, carried a crew of six, and had two hosereels, mounted one on either side. An overhead gantry carried two sets of extension ladders, and tools and hydrant fittings were carried in the running board lockers, with the hard suction hose strapped to the one on the left side.

A few years ago a similar machine was discovered in Denmark and driven to the Scania museum at Malmo where another interesting machine is on show – a 1919 4 × 4 Scania Vabis fire appliance.

Fig. G American La France-Hibernia aerial ladder (USA)

McLaughlin Airfield Tender   *Plate 50*

The earliest airfield tenders were conversions of proprietary truck chassis embodying equipment found on the normal town and country engines. Such vehicles were capable of attacking the fires arising out of aircraft crashes and the first consideration was to douse the flames just enough to enable a rescue of the crew to be made. The aircraft were largely of wood and canvas construction with small fuel tanks and a crew of one or two.

A vehicle of this category is the McLaughlin airfield tender pictured at the RAF airfield at Camp Borden in Canada at the end of World War I (plate 50). This particular machine was one of the first type to be used by the RAF and carried chemical foam in two large cylinders at the rear with a small-bore hosebox on top. A short extension ladder and rolls of fire hose appear to have been carried anywhere that was convenient with nozzles and tools in running board boxes.

The provision of snow chains on the rear tyres and a tarpaulin thrown over the bonnet remind us that the machine had to stand ready to turn out in all weathers!

Ford Model T   *Plates 51–2*

With a total production figure of 15,007,003 for the Ford Model T it is not surprising that many fire engines in the United States and elsewhere used this chassis or at least embodied some of its parts. It was used in many different roles. In some instances it was merely used as transport for the crew, in others it was a commercial truck with certain fire equipment added, whilst on very many occasions it was turned out as a fire truck by specialist firms.

One such specialist builder was the Howe Fire Apparatus Company of Anderson, Indiana who supplied pumpers, chemical engines, ladder trucks and combination equipment on proprietary chassis.

The vehicle illustrated in plate 52 is the Howe Model No. 4 chemical and ladder equipment mounted on a Ford one-ton (TT) chassis. This particular machine was supplied to the Trufant Fire Department in 1927 and was equipped with two 40-gal. soda acid type chemical tanks, the contents of which were fed into a 150-ft hose reel mounted above them. An extension ladder with folding hooks was carried on the left side while a stack of metal buckets, a funnel and portable extinguishers were carried on the right. A well-upholstered bench seat was supplied for the driver and officer while the rest of the crew were catered for by the wide step and large grab rail at the rear. A box for fittings and tools was provided behind the bench seat and a fire siren was provided to clear a way through the traffic.

Plate 51 shows an early Ford Model T converted for use on railroad track for extinguishing fires in the grass and bush adjoining the right-of-way. It will be noted that this vehicle carries the simplest of equipment – shovels, rakes, beaters and a sand bin at the rear end. It was essentially a first-aid machine and if the crew came across a serious fire they just made straight for the nearest depot for help!

Mack Fire Engines *Plates 53–5*

Starting in a wagon shop in Brooklyn, New York, in 1900 the five Mack brothers set up business producing truck and bus chassis. After five years the business was moved to Allentown, Pennsylvania, which is still the site of part of Mack vehicle production.

In 1909 a Mack tractor was supplied to the local fire department for towing a 75-ft aerial ladder and this is the first recorded sale of a truck for fire use. In 1911 the first Mack fire apparatus proper was delivered, a combination chemical and hose appliance with midships pump.

A long-wheelbase ladder truck was produced in 1910 and the first left-hand-drive fire apparatus was produced in 1915. Mack also claim the first fire truck with four wheel brakes in 1927, and engine-driven aerial ladder in 1928, a six-wheeled apparatus in 1935 and an enclosed appliance in 1936.

Returning to 1911 we find that Mack joined forces with two other motor truck builders, the Saurer Motor Company (of Swiss origin) and the Hewitt Motor Company. This consortium was called the International Motor Company.

Synonymous with the Mack name is the rather ugly AC model front end which was introduced in 1915. This style of truck gained an enviable reputation for toughness and stamina as a result of its use by the American forces in Europe during the First World War. As a term of endearment they were known as 'Bulldog Macks' to the forces and that name stuck forever, the company insignia being a bulldog and that being the name of their house journal today. The style continued until 1939.

During World War II Mack produced approximately 200 special airfield crash tenders for the US Army. These were CO_2 dispensing vehicles based on 6 × 6 chassis and carried 1,000 gals of water.

In the post-war years Mack have produced pumpers, hose trucks, ladder trucks, triple combinations, quadruple combinations, articulated aerial ladders, aircraft crash tenders and special rescue trucks.

In plate 53 a Mack model AC tractor of about 1925 is shown coupled to an old horse-drawn water tower which has been adapted for motor traction to extend its useful life. Plate 54 shows another AC model tractor of about the same period but this time coupled to a 75-ft aerial ladder with rear steering. The facility of rear wheel steering on the semi-trailer enabled the long articulated outfit to negotiate street corners more easily.

The third AC model Mack is a mid-1930 pumper and hose car with twin hosereels (plate 55).

Motorcycle Combination Appliances *Plates 56, 58*
In the Merryweather & Sons catalogue for 1895 was a quadricycle hose carrier machine consisting of two bicycles suitably connected with tubes and with a hose box positioned between them. The design was said to enable the officer and three men to cycle ahead of the horsedrawn pump, find a hydrant and proceed to connect the standpipe and hose in readiness for the arrival of the pump.

In 1910 a further design of light machine was introduced when the first Merryweather motorcycle combination appliance was evolved. This outfit was a truly 'first aid' machine with a small manual pump carried on the side-car frame.

A few years later another type of machine was introduced, this time with a box-type side-car capable of holding 500 ft of hose together with standpipes, branch-pipes, breeching and tools. The rear of the side-car box was provided with a roller so that once the hose was connected to the hydrant standpipe the machine could be driven forward and the hose payed out on the move.

Another type of Merryweather machine incorporated the AJS twin with a 20-gal. chemical extinguisher and 40 ft of small-bore hose carried on the side-car chassis. This design was said to have been developed for use in China.

The motorcycle combination illustrated in plate 56 is a Leyland production based on a specially adapted BSA V-twin machine. A small self-contained pumping unit together with hard suction hose, rolled delivery hose, branch-pipes and breechings are all carried on the side-car chassis, while the crew of two are carried on the motorcycle.

An Eastern European design for a similar type of machine is shown in plate 58.

Obenchain and Boyer *Plate 57*
The origins of this organisation go back as far as 1875 when Messrs Obenchain and Boyer opened a mill. After they had moved to a new site in 1897 the premises were destroyed by fire. It may have been more than a coincidence that as soon as the premises were rebuilt we

find that Mr Obenchain obtained a patent for a new type of chemical fire extinguisher.

The new extinguishers soon gained a ready market and sales were made in all parts of the United States and Canada as well as to Australia, China, South America and the West Indies. Production of hand-propelled chemical engines began in 1905. The company suffered a major setback in 1909 in the sudden death of its leader. 1913 saw the company incorporated as the Boyer Fire Apparatus Company Inc. and in 1916 the first motor-driven appliance was produced by taking a Ford Model T roadster and adding a chemical tank and short ladders. Many engines were to follow this pioneer and be mounted on Ford chassis.

Business boomed and in 1917 the company was recapitalised and moved to larger premises in Park Street, Logansport, Indiana. This put the company into a better position to meet the commitments of World War I, in the course of which they produced several hundred hand-drawn chemical engines. Similarly, after America entered World War II many hundreds of motor apparatus were supplied to the US Army. The last change in company control was in 1952 when under the management of Mr Harry Armington Boyer became Universal Fire Apparatus.

As with many other specialist fire apparatus builders a variety of chassis have been used for the mounting of equipment, and Boyer bodies have been seen on Ford, International, Packard, Dependable, Clydesdale, Dodge, Diamond T, GMC, Chevrolet, Studebaker, Mack, Federal and White chassis. The vehicle depicted in plate 57 is a Clydesdale of around 1920.

Teudloff Dittrich *Plate 58*

The greatest and oldest Hungarian fire apparatus firm was undoubtedly the Teudloff Dittrich concern of Budapest/Kispest, which was founded in 1885. Starting with hand fire pumps and horse-drawn pumps the company later went in for building steam fire pumps. As the petrol engine became more evident as a power unit it was used to power fire pumps, being utilised for trailer, horse-drawn and motorcycle applications as well as for the more orthodox motor chassis types of fire engine. The company were successful in obtaining many patents relating to the manufacture of motor fire appliances and during the 1920s increased their range of equipment. Pumps of up to 2,600 litres per minute output were produced and vehicles were turned out with pumps mounted in front of the radiator, midships or at the rear.

Naturally, as the company did not produce vehicle chassis as such it used to mount the equipment on chassis specified by the customer, and appliances left the works mounted on Raba, Raba-Austro-Fiat, Raba-Krupp, Mag, Hungarian Fiat, Berliet, Bussing, Minerva, Mercedes-Benz, Austro-Fiat, Mavag, Mavag-Mercedes-Benz and Somua chassis as well as on Meray motorcycles.

As well as fire engines the company produced all kinds of other municipal vehicles and some of these were so constructed as to be capable of fire fighting if so required. Teudloff Dittrich amalgamated with the Mavag concern in 1932 to become the Royal and States Iron, Steel and Machinery Works, Budapest/Kobanyai. During its life of some seventy-two years vehicles were exported to a total of forty-five countries.

The appliance shown in plate 58 is a Teudloff Dittrich combination appliance utilising a Meray motorcycle. This type of machine was produced with a JAP 500 or 600 c.c. side-valve engine and had a self-contained demountable engine/pump unit installed on the side-car frame. Two lengths of suction, four lengths of delivery hose and two branches were also carried on the side-car. It is interesting to note that the side-car is positioned on the left hand side of the machine.

Commer Canteen Van *Plate 59*

At large fires where a great number of firemen are working for long periods it is important that they are relieved whenever possible so that fatigue does not render them susceptible to accident. Similarly they require adequate refreshment if they cannot be released from duty for very long, and mobile kitchens and canteens are necessary to keep the crews supplied with food and drink.

The vehicle shown in plate 59 is a canteen van which was used by the London Fire Brigade for many years. The chassis was a 30–36-h.p. Commer which first saw service in 1908 when it formed the basis of a hose tender carrying a short extension ladder and a crew of six. Some years later when the hose tender was replaced with a more modern machine its body was removed and replaced by the one shown in our illustration. This body in fact was originally a horse-drawn canteen built around the turn of the century by the Belle Isle Company. It was similar in design to the ordinary coffee stall which has now almost disappeared from the street corners of large cities.

Catering equipment consisted of boilers, cupboards, a dresser, crockery and cutlery necessary for dispensing coffee, cocoa, Oxo, biscuits and cheese.

This vehicle lasted until 1935 when it was replaced by the more modern canteen shown in plate 95.

Delahaye, 1926

One of the earliest French automobile manufacturers, Delahaye has origins going back to 1896. Before production ceased around 1956 the company had been responsible for building many fire appliances for the Regiment des Sapeurs-Pompiers de Paris and for the Corps de Pompiers in many other French towns, as well as supplying vehicles for use in the French colonies.

Toward the end of the 1920s it was reported that Delahaye were responsible for about ninety per cent of the engines used by the Corps de Sapeurs Pompiers. At this time they built three standard types of appliance for use in Paris. The 1926 range consisted of open machines with large water tanks, rear-mounted pumps and detachable hosereels, with outputs from 60,000 to 110,000 litres per hour. Another type of open machine used the maker's 12-h.p. (French rating) large touring car chassis and cleverly adapted it so that as well as being equipped with a midships pump, hook ladder, water tank, hosereel and detachable hosereel it still managed to find space for a crew of five, three of whom sat across the rear facing backwards! It speaks highly of Delahaye workmanship when one considers that it was still capable of 50 m.p.h. (Plate 60.) The largest vehicle in the range was the 30-h.p. 30-metre mechanical turntable ladder which at this time (1926) was the only remaining chassis sold on solid rubber tyres, the remainder of the range being available on pneumatics.

Probably the most unusual vehicle of the period was the totally enclosed pump for the Paris fire fighters. This machine provided covered accommodation for a crew of fourteen as well as carrying a powerful pump in the rear compartment which was capable of an output of 120,000 litres per hour. The usual detachable hosereel was carried at the rear.

Laffly, 1926

The Laffly concern had its beginnings in 1922 and during the period up to 1939 was producing light vans, lorries, buses, agricultural tractors, road rollers, municipal vehicles and fire appliances.

One type built is depicted in plate 61. This shows a light fire appliance named the Auto-Pompe Hydro-Chimique or chemical engine, which used the Mousse system of foam emulsion. The chassis used was the maker's 2-ton LC model which had a four-cylinder petrol engine of about 20 h.p. driving through a multi-disc clutch and four-speed gearbox to a bevel gear rear axle. Fire-fighting equipment consisted of a midships-mounted centrifugal pump with hose connections at the rear and controls at the front. Water storage amounted to

800 litres in two tanks, and by the addition of only 6 kilos of Mousse foam compound it was asserted that 50,000 litres of foam was produced. It is interesting to note that the two water tanks were fitted with hand-operated agitators as initial foaming agents!

Three lengths of suction hose were carried on the side running boards, about fourteen rolls of hose were carried in a cage on top of the water tanks, and a small-bore hosereel was fitted transversely just below the rear ladder gallows. A short extension ladder and six hose branches and nozzles made up the rest of the equipment. Six of the crew members were seated on two cross benches at the front, while the remaining four had to sit on the water tanks!

The complete vehicle measured 5 m. long, 2 m. wide and 2 m. 40 cm high and weighed approximately 3,000 kg.

Scemia, 1927 *Plate 62*

A type of fire-fighting vehicle which has had many applications on the continent of Europe was really not a fire engine in the strict sense of the word at all. It was in fact a municipal street washer but fitted with certain additional equipment in order to render it capable of carrying out fire-fighting duties when so required. Most municipal street washers consisted of a water tank and means of spraying the water on to the road surface. The water tank was refilled from hydrants situated at intervals along the roadside. On some types a power-driven broom was added in order to sweep refuse to one side as the vehicle was driven forward. Others had a water pump fitted so that the water could be ejected in a more powerful stream and sometimes a hosereel was carried in order to wash down areas such as market places where the vehicle could not go.

So it was just a short step to making the small town street sprinkler into a full fire-fighting vehicle in everything but speed – vehicles designed as street sweepers and washers are usually geared for slow speeds.

A French version of a vehicle of this type is the Scemia shown in plate 62. Basically a street sprinkler/sweeper with a 3,200-litre water tank, it carried a rear-mounted centrifugal pump with a capacity of 72,000 litres per hour and a hose reel with 60 metres of small-bore hose. Four lengths of suction were carried on top of the tank, surrounded by a handrail which was used as security for the crew of twelve standing on a narrow platform on either side of the water tank. The vehicle had a wheelbase of 3,000 mm. and an overall length of 6,200 mm., stood 2,550 mm. overall and weighed 5,100 kilos unladen. Power was supplied by a four-cylinder engine of 100-mm. bore and 160-mm. stroke, producing 47 h.p. at 1,500 r.p.m.

Ford-Howe Pumping Chemical and Hose Combination   *Plate 63*

Following the outstanding and successful long production run of the Model T Ford, the introduction of the Model A in 1927 was soon seized upon by Ford dealers and users anxious to put the new model through its paces. Numbered among the thousands of users of the ubiquitous Model T were many Fire Departments and they, no less than the multitude of deliverymen, hauliers, dairymen, farmers, road-builders and others, were anxious to see the Model A and try it for themselves.

One such a test was carried out by the National Board of Fire Underwriters on May 11th 1928 when their Committee on Fire Prevention and Engineering Standards took a Howe-Ford Model HBS Pumping Chemical and Hose Combination for pumping tests (plate 63).

The particular fire truck put to the test was Model A No. 92896, fitted with a four-cylinder gasoline engine rated at 24 h.p. (ALAM rating) which produced 40·5 b.h.p. at 2,000 r.p.m. A Waterous Model B pump was fitted to the machine and other equipment included an 80-gal. water tank mounted transversely behind the driver's bench seat, foam-making equipment, hose reel, buckets, extension ladder, portable extinguishers, hooks, axes, nozzles, suctions, searchlight and siren.

The desired performance of the pump was to produce 300 gals per minute at 120 lbs pressure, 150 g.p.m. at 200 lbs pressure, and 100 g.p.m. at 250 lbs pressure. Taking water from a nearby creek with a lift of 4 ft the pump gave a good account of itself and the following results were recorded. In the first test, of six hours' duration, pumping through 150 ft of $2\frac{1}{2}$-in. hose with $1\frac{1}{8}$-in. nozzle, using third speed (2.5:1 ratio), the engine turning at 1,884 r.p.m. and the pump at 738 r.p.m., the average rate of discharge was 318 g.p.m. at 128 lbs pressure. The second and third tests both lasted three hours and in both 200 ft of $2\frac{1}{2}$-in. hose with $\frac{3}{4}$-in. nozzle was used. In the second, the engine turning at 1,876 r.p.m. in second gear (4.63:1 ratio) and the pump at 405 r.p.m. produced a discharge of 159 g.p.m. at 207 lbs pressure. In the third the pump gave 110 g.p.m. at 258 lbs pressure, at 318 r.p.m. (engine speed: 1,476 r.p.m.). In his concluding remarks the Engineer carrying out the tests reported that the tests indicated that both the engine and the pump had considerable reserve capacity.

Raba Fire Engines   *Plates 64–5*

The Hungarian Railway Carriage and Machine Works Ltd, of Gyor, Hungary, was established in 1896 and continues production to this day. In 1904 an Automobile Works was established under the control of the railway carriage works and production of pleasure car chassis for the Spitz works in Vienna was begun.

Around 1910 production of lorries commenced and a few years later Raba-Austro-Fiat trucks were produced under licence from the Austro-Fiat concern. Shortly afterwards Raba-Krupp lorries appeared after an agreement was reached with the Fried Krupp works of Germany.

From 1916 buses, emergency vehicles and fire engines were built on chassis designed for $2\frac{1}{2}$ to 5 ton payloads. The smaller fire appliances were based on Austro-Fiat designed 2-ton chassis with 40-h.p. engines, 1,600-litre water tank, and two-stage centrifugal pumps with an output of 800 litres at 6 atmospheres pressure.

Larger fire vehicles were mounted on 3 to 5 ton payload chassis of Krupp design, and this range was fitted with engines in the 50–70 h.p. range, water tanks of 2,000 to 5,000 litres capacity and pumps producing 1,200 or 2,000 litres per minute.

The works was nationalised in 1948 and fire engine manufacture ceased, although in 1967 vehicle production recommenced with an agreement to produce vehicles under licence from MAN.

Two vehicles from the Hungarian works are illustrated, the first being a 1916 Raba-Krupp chassis with rear-mounted pump, twin hose reels, detachable devidor hose reel and some vicious-looking hook ladders! (Plate 64.) The second machine is of the 1920s period and is a lighter vehicle based on the Raba-Austro-Fiat 2-ton chassis with 40-h.p. engine, 1,600-litre water tank and two-stage centrifugal pump of 800 litres per minute output (plate 65).

Rosenbauer Appliances *Plates 66–7*

Johann Rosenbauer founded the business which bears his name in 1866 at Linz in Austria. In common with other fire engine builders of the period he produced manual and horse-drawn equipment and extinguishers and it was not until 1908 that he built his first motor-driven centrifugal pump.

In 1910 a small demountable pump was perfected, and in 1923 the first portable fire pump powered by an air-cooled two-stroke engine was available. 1925 saw the first Rosenbauer fire engine with pump mounted on a front extension of the engine crankshaft, and later in the same year a portable pump driven by a four-stroke engine was designed.

The two appliances shown in plates 66 and 67 are Rosenbauer-equipped machines mounted on Steyr private car chassis of the 1929 era. During the 1930s Austro-Daimler, Perl, Austro-Fiat and WAF chassis were used for mounting fire appliance bodywork as well as the regular Steyr chassis.

Plate 66 shows a vehicle with front-mounted pump, two seats for the crew and racks for rolled hose above a rear compartment for tools and equipment. The hand-operated ladder truck shown in plate 67 has an extension to 14 m. only, and the vehicle carries no pump.

Magomobil *Plate 68*

The vehicle illustrated in plate 68 shows a personnel carrier rather than the type of vehicle normally associated with fire brigades which are primarily carriers of equipment with associated crew.

Often brigades supply a car for use by the officer in charge, and in some instances the brigades have seized this opportunity to provide speedy transport for the movement of additional or relief crews when required. Basically a touring car chassis, possibly modified with regard to road springs for the heavier load, this vehicle has had the rear end of the body modified in order to provide a double cross bench seat arrangement utilising a common leg well for compactness. In order to make the vehicle even more useful once it has carried out the task of bringing personnel to the fire scene, a front-mounted pump is fitted. This has direct drive from the engine. Running-board-style lockers carry hydrant fitting and other tools and a small hose reel is fitted at the rear.

The centrifugal pump capable of pumping 500 litres per minute at a pressure of 7·5 atmospheres, together with the other fire-fighting equipment, was supplied by Teudloff Dittrich of Budapest/Kispest on a Magomobil chassis produced by the Hungarian General Machinery Works of Budapest/Matyasfold. Many fire engines on MAG chassis were produced during the period 1922 to 1935 and they saw service with many brigades and performed well although they were of moderate proportions compared with the larger machines produced by Teudloff Dittrich on strictly commercial vehicle chassis.

Ahrens-Fox Fire Engines *Plates 69–71*

In 1908 the Ahrens Fire Engine Company was formed in Cincinnati under the direction of John Ahrens. A little later Charles Fox joined the company and it was re-organised to become the Ahrens-Fox Fire Engine Company.

The first product of the company was the Ahrens Continental steam fire pump, but as the motor vehicle was developing so rapidly at this time, Ahrens-Fox were keen to produce a petrol-driven pump which combined the pumping power of the steamer with the speed of the petrol motor.

In 1911 the first petrol-driven Ahrens-Fox machine was produced – the Model A. This was a six-cylinder ($5\frac{3}{4}$ in. $\times$ $6\frac{1}{2}$ in.) machine with a two-cylinder piston pump producing 750 gals per min. at 120 lbs per sq.

in. The style of the machine is legendary in fire engine history – the pump mounted far out in front of the vehicle radiator between the front wheels with large air vessels and water cooler above. This water cooler was found necessary because during sustained pumping the normal radiator was unable to provide enough cooling capacity.

During the next few years Ahrens-Fox fire apparatus received favourable comment and many machines entered service both in the USA and abroad. In addition to pumpers, which by now were sporting a re-designed spherical air chamber, ladder trucks, hose cars, combination trucks and articulated tractors for old horse-drawn ladder trucks were turned out.

In 1923 a new type of apparatus in the shape of an 85-ft aerial ladder was produced, this outfit being equipped with a small engine-driven compressor to provide power to elevate the ladder. In 1927 a 1,000-gal.-per-min. pumper was produced and in 1929 the Tower Aerial was introduced which gave double reinforced tubular truss rods on the ladder.

In the early 1930s business was poor because of the slump and few apparatuses were sold. In a bid to stabilise their company Ahrens-Fox merged with the Le Blond-Schacht Truck Company in 1936.

1939 saw the company favoured with an order for five pieces of apparatus to protect the New York World's Fair. All these five vehicles were acquired by the New York Fire Department when the Fair closed.

In 1951 the company passed into the hands of Walter Walkenhorst and Ahrens-Fox equipment began to appear on proprietary chassis in an effort to gain orders from small municipalities. 1953 saw the start of an agreement with C. D. Beck and then in 1956 control passed to Mack Trucks. Fire trucks under the Mack name continued for a short while from the old Beck premises and then transferred to Allentown, Pennsylvania.

Three Ahrens-Fox machines are shown in plates 69–71. The first (plate 69) is a 1924 650-gal. pumper and hose car which shows the typical style of the front-mounted pump and spherical air chamber. Next is one of the five machines produced for the New York World's Fair (plate 70), and last is an apparatus in a different style, a 1939 Pumper and Hose Car with a midships-mounted 1,000-gal.-per-min. pump (plate 71).

Fire Engine Equipment *Plates 72–3*

One of the attractions of fire engines to an enthusiast is the fact that no two are exactly alike. Looking back in time there has been a most varied assortment of designs with innumerable differences in chassis type, propulsion, type of pump, style of bodywork, length of ladder, size of engine, disposition of crew and name of builder.

There have been moves toward standardisation with regard to some items of equipment in some countries, and this will no doubt become more widespread as time goes on. There is no doubt that a standardised layout of equipment in a particular brigade does help in the training of appliance crews and brings greater efficiency.

Returning to the appearance of the appliances, it is hoped that the reader will notice the great variety of equipment shown and described in these pages although there are plenty of other designs worthy of attention that have had to be passed over. To the fire engine buff or enthusiast the appearance of the machine is the thing which has a lasting effect. Some machines look decidedly fussy (plate 64) while another has a distinctively powerful appearance (plate 71). Obviously the proportions of the complete vehicle have a bearing on its appearance and to compare the three-wheeled Scammell (plate 99) with the six-wheeled Range Rover (plate 167) is decidedly unfair.

However interesting these comparisons may be we should not forget the function of the vehicle in connection with its age. As mentioned elsewhere, at one period in fire engine history the Braidwood body style (plate 74) reigned supreme because of its simplicity and function merely as a carrier of men and equipment. Then, within a few short years, the whole aspect of fire engines was turned inside out by the demand for an improved design and the enclosed or limousine appliance became all the rage (plate 92).

Fig. H Reo combination, 1935 (USA)

Another factor concerning the overall appearance of the machine is the very equipment it has to carry, and the area in which it operates. Plate 72 depicts a machine about as bare as one could get – a simple Braidwood body with seats for the crew on boxes for the hose and a ladder carried aloft. True it is not really a fire engine at this stage because the towing eye tells us that it normally tows a trailer pump. Now compare it with the Reo for Shanghai illustrated in fig. H, with its enormous load of equipment. One wonders if the front wheels came off the ground while the machine was climbing a hill and, more important, how the crew managed to cling on and stay alive.

A smaller machine of the same brigade is shown in plate 73, and although this is an earlier type than the Reo it shows the crew to be in a safer position inboard of the machine. The extra-long suctions seem to enclose the machine like giant cobras but any smile on our faces is soon removed by the sight of the crew in gas masks reminding us that being a fireman in Shanghai was not all that funny.

Ford-Simonis *Plate 74*

'The requirements for fire protection of smaller municipalities, villages, country districts, estates, etc., call for a fire engine easy to manipulate, requiring a minimum of training, fast, reliable, light, so that it can negotiate country roads, and particularly at moderate initial cost and low cost of upkeep . . .'

So went the opening paragraph of a leaflet describing the 'new' Ford chassis with Bateman body and Simonis equipment introduced at the end of 1931. (Plate 74.)

The Model AA chassis with 131-in. wheelbase and 24-h.p. engine formed the basis of this machine which, as the publicity announced, was aimed at the less wealthy rural brigades which at this time were served by a motley selection of equipment including in some instances old horse-drawn steam pumps towed to the scene of the fire by a council lorry! Bodywork consisted of the normal Braidwood-style body with the minimum of comfort for the crew of driver, officer and six men. The box body was provided with side doors for the storage of hose and running board boxes were available for the storage of standpipes, hydrant tools, etc. Although hose and hydrant fittings were listed as extras the machine did come complete with 25 ft of patent composition suction hose and two strainers. A 25-ft extension ladder by Simonis Limited was carried on a ladder frame and the patent self-priming Simonis turbine pump was mounted at the rear. A 25-gal. water tank and 120-ft first aid hose reel was available at extra cost if required.

Bedford Tender, 1936   *Plate 75*

Towards the end of 1936 the fire brigade of Uxbridge in Middlesex took delivery of what was described at the time as a 'first-aid saloon tender'. This machine was designed as a first-call appliance and for small fires, and as such was built on light lines in order to be capable of a quick turnout (plate 75).

Based on the Bedford 2-ton chassis, the appliance was fitted with the standard Bedford six-cylinder petrol engine which in addition to powering the vehicle was used to drive the Winn pump by means of a power take-off from the vehicle gearbox. Behind the driver was a 50-gal. water tank with connections to a 250-ft hose reel which was mounted on top of the water tank.

Bodywork on the vehicle was supplied by Gregorys of Uxbridge and the design provided completely covered accommodation for the crew who sat on seats at the sides of the body. These seats also doubled as lockers for the loose equipment, the contents of which could be reached from outside the vehicle by means of hinged flaps. Rolled hose, branches, standpipes, nozzles, ropes, axes, pick-axes and shovels were all carried in these side lockers. Access to the body itself was by means of a hinged door on each side of the body just behind the driving cab, and it was through these doors that the hose reel was fed when in use. At the rear of the body was a wide vertical roller shutter designed to give ample room for the swift exit of the crew.

A 30-ft wooden extension ladder was carried on the roof, and hydrant connections for the Winn pump were located on the nearside of the vehicle together with gauges and controls for its operation.

In later years this vehicle saw service with the National Fire Service and Middlesex Fire Brigade. It ended its fire service days as a mobile canteen.

Air Force Fire Vehicles   *Plates 76–87*

Plates 76 to 87 illustrate some of the types of fire-fighting vehicles and equipment used by the Royal Air Force during the period 1930 to 1945, in an endeavour to show how appliances changed during that time. Other types used by the Air Force are shown in plates 50 (McLaughlin), 133 (Thornycroft) and 119 (Alvis Airfield Crash Tender).

During 1930 the Merryweather Hatfield pump shown in plate 76 was introduced to RAF service and additional equipment for aircraft fires was added. A 30-gal. foam tank was mounted behind the front seats and the discharge hose for this was stored in the hose box between the tank and the pump. For accurate positioning of the foam the discharge end of the hose was fitted with a long metal tube with funnel-shaped end. The long suction hose was carried permanently connected to the pump and curled round the four 2-gal. extinguishers at the rear and continued along the nearside of the vehicle. A long extension ladder was carried on overhead gallows and running board boxes were provided for tools and fittings.

Plate 77 shows a type of vehicle much more suited to airfield fire fighting duties than the type described above. Based on a six-wheeled Morris Commercial chassis which gave a better cross country performance, the appliance was designed for fighting aircraft fires and was not an adaptation of a civilian machine. Equipment included three 30-gal. chemical foam generators, with methyl bromide additive as an attempt to improve the flame suppression power of the foam, and twenty 2-gal. foam extinguishers. The output from the three large foam generators was fed through long hoselines which included rigid pipes for better positioning on the fire.

In 1932 the National Fire Protection Company demonstrated the little appliance shown in plate 78. This was a conversion of a Triumph Nine car equipped with two banks of five chemical extinguishers manifolded into a mixing tank and then piped to a hose reel. A small engine-driven pump was fitted under the seat and a three-section extension ladder and preventer were carried on overhead gallows.

Also in service about this time was the vehicle and trailer shown in plate 79. The truck carried a 50-gal. foam tank with CO_2 pressure discharge and four 60-lb CO_2 cylinders discharging through a hose reel with the funnel-shaped applicator. The two-wheeled trailer was similarly equipped with CO_2 cylinders and hose reel.

During 1936 the streamlined Crossley tender shown in plate 80 was introduced. It was based on a 6×4 chassis layout. This maker was to supply many fire-fighting vehicles to the armed forces during the 1939–45 war.

This particular design carried a 200-gal. water tank, foam equipment and four 80-lb CO_2 cylinders. The saponine foam was discharged by two airfoam pumps which were chain driven from the vehicle engine.

Another style of six-wheel vehicle current about this time was the Ford BB model CO_2 and water tender shown in plate 81. Four hose reels were provided on the appliance, one for water discharge and three connected to the bank of 60-lb CO_2 cylinders positioned over the rear bogie.

Just before the start of World War II the Crossley FE1 type of appliance was introduced into service with the Air Force (plate 82). Based on the manufacturer's 6×4 chassis, this appliance was equipped with a 300-gal. water tank, 28-gal. foam liquid tank, four 60-lb CO_2 cylinders and a positive displacement pump.

Plate 83 shows a Karrier Bantam platform lorry with chemical foam tank, ladders and hoses for fighting small fires on Air Force property. This type of appliance was not intended as an aircraft fire-fighting vehicle.

The next example is a six-wheeled Ford tender based on the war-time WOT1 chassis, equipped with a 400-gal. water tank and 28-gal. foam liquid tank. The airfoam pump is housed in the enclosed compartment at the rear and discharges through side hose connections. This appliance has no hose reels or CO_2 cylinders for fire fighting. (Plate 84.)

A later version of the Ford WOT1 type is shown in plate 85. This type has a 300-gal. water tank, a 100-gal. foam liquid tank, an airfoam pump powered by a separate Ford V8 engine and three rearward-facing monitors. Two of these monitors are mounted on the rear platform whilst the third is mounted on top of the monitor tower which can be folded down during transit. Another feature of this type was that at prolonged fires it could be kept supplied with water from a separate water tender.

Plate 86 shows a local conversion of a US Jeep into a small rescue truck. A short ladder, fire extinguishers and tools were carried on the vehicle for use during rescue operations.

As an experiment in 1945 a captured German half-track by Mercedes Benz was converted into a fire appliance using water and foam tanks, with separate engine for the pump and rear-mounted monitor. It is reported that the project was not pursued because of damage to perimeter roads and runways! (Plate 87.)

Six-wheel Appliances *Plates 88–90*
In the years following World War I the motor vehicle showed itself to be capable of carrying out most of the transport tasks asked of it. The light and medium aspects of goods haulage, the majority of bus work and the greater portion of fire engines built were powered by the petrol engine. It was only in the heavy section of goods haulage that steam persisted.

The exploitation of colonies demanded transport systems and although railways were built there were many thousands of miles where road transport played a vital role although often there were no roads as such. These untamed areas required tough vehicles capable of taking great punishment and many vehicle manufacturers produced models of the 'colonial' or 'cross-country' type to open up the new territories.

Guy, Morris-Commercial, Thornycroft, Leyland, Garner and others turned out six-wheel chassis embodying various types of suspension in order to achieve maximum traction over unmade roads or across open country. Naturally some of these chassis were adopted by fire engine builders as the basis for appliances requiring cross-country ability to maintain fire protection in country areas. In some cases the abilities of the vehicle were further enhanced by the fitting of detachable tracks around the wheels of the rear axles so as to provide the vehicle with traction something akin to a military half-track. In fact some of the six-wheel designs were based on chassis coming with the scope of the government specification for military vehicles. There have been a few fire appliances based on true half-tracked vehicles but these have distinct disadvantages when they are used on metalled roads.

Another aspect of the six-wheeler was the greater body length obtained with some designs. During the 1920s and 1930s fire appliances were being asked to carry increasing amounts of equipment and having reached the limit on two axles a third was added, in either rigid or articulated designs.

There are several six-wheel appliances illustrated in this book, some where the three axles are used for a greater length of vehicle, and others where traction over open country has been the deciding factor. Fig. Q shows a six-wheel layout used to give support to a long hydraulic platform, while in plate 90 is a design for cross-country work.

The Thornycroft-Simonis appliance shown in plate 88 is a type based on a War Office subsidy model chassis of 1931 and is showing the flexibility of its rear bogie as it crosses undulating country.

In plate 89 is shown a six-wheel version of the Leyland Cub supplied with detachable tracks for the rear axles. This 400-gal.-per-min. pump with New World style body was produced in 1933.

Other six-wheel appliances with a variety of body styles will be found in plates 80, 81, 85, 91, 115, 119, 130, 136, 155, and 157.

Leyland Tender for Speke Airport　　　　　　　　　　　*Plate 91*
Early in 1939 this Leyland six-wheel airfield crash tender was commissioned by Liverpool Fire Brigade primarily for the City Airport at Speke. Registered FKA 844, this appliance was based on the well-known Cub chassis with a six-wheel layout which incorporated a double-drive rear bogie which was fully articulated. Designated KZDX 2 chassis type, this eight-speed machine utilised 9.75×20 'trakgrip' tyres on each of the six wheels and carried bodywork by Leyland Motors with special fire-fighting equipment by Pyrene. Crew

accommodation consisted of double cross-bench seats behind a two-piece toughened-glass windscreen. Over the rear bogie was mounted a 650-gal., bitumen-lined water and foam tank the contents of which were pumped by the 500/700-gals-per-min. centrally-mounted two-stage centrifugal pump, through any or all of the four longitudinally-mounted hose reels each containing 100 ft of $1\frac{1}{4}$-in. diameter non-kinkable hose. Each of the hoses terminated in a Pyrene foam gun and each could produce foam at 600 g.p.m.

Designed by Chief Officer George S. Oakes of Liverpool Fire Brigade for use at the expanding Speke Airport, the appliance could be used for normal fire-fighting duties, and for this purpose a 4-in. suction connection was provided at the rear of the chassis.

Enclosed Appliances *Plate 92*

From the days of the horse-drawn manual engines right up to 1928 the facilities for seating the crew on British fire engines had changed little, if at all. During the thirty years that James Braidwood was in control of the London Fire Engine Establishment he did much to co-ordinate the twelve fire insurance brigades into a single combined fire-fighting force and one of the ways in which he is remembered is by the seats-on-top-of-hose-box design of bodywork which acquired the title of Braidwood body. This was because of his idea of designing a standard type of appliance for the London brigade and the general layout found favour with other brigades and followed into the motor engine era.

As the speed of engines increased so the fireman's task of keeping on the engine as it thundered along to the fire, weaving through traffic and sliding round corners, became more difficult.

The fear of the firemen and the anger of the public increased as the number of deaths and injuries to firemen steadily went on, and the death of Fireman Kidd at Chippenham in 1929 brought the matter to a head. *Fire* magazine declared that it was time for fire appliance designers to turn their attention to the protection of the crew because the Braidwood-style body did not offer protection to the crew of a fast-moving motor appliance.

It was about this time that attention was being turned to a design of body that had originated in America and found use in certain European countries. This was described as the 'inside type' because of the way in which the crew sat in two rows facing each other inside the body of the machine. One of the first machines to appear with this New World style of body was a Dennis with midships-mounted pump built to the design of Chief Officer A. Andrew for the Luton brigade. Birmingham Fire Brigade followed suit acquiring the new-style body on Dennis and Leyland chassis using a driver-alongside-engine layout in order to reduce overall length.

Soon afterwards, in 1931, Edinburgh and Darlington took delivery of totally enclosed appliances and during the 1930s many designs of enclosed or limousine machines were to join the fleets of brigades throughout the country.

The enclosed appliance is a rarity in America but it has found great favour in all parts of Europe.

The vehicle illustrated in plate 92 is a 1935 Triangel of Copenhagen Fire Brigade, equipped with Hercules engine and 1,600-litres-per-min. rear-mounted pump.

A vehicle with the New World style of body is the Leyland six-wheel appliance shown in plate 89. Vehicles with the older Braidwood-style bodywork are shown in plates 47, 64, 72, 74, 97 and 98.

100-ft Turntable Ladder   *Plate 93*
Looking back over the history of fire appliances one is struck by the amount of publicity given to the turntable ladder. Admitted it is a fine piece of engineering in principle and execution, and a valuable asset to any brigade particularly in the role of rescue appliance, but the advertisements of various turntable ladder builders at one time made one think that no other piece of apparatus could claim so many attributes.

We have been treated to mechanical ladders, hydraulic ladders, wooden ladders, steel ladders, lightweight ladders, ladders with monitors, ladders with pumps, ladders with cages. In fact over the years it seems that no other item of equipment has caught the imagination quite like the turntable ladder. Probably no other period in fire engine history was quite so colourful as the eleven years from 1927 to 1938 when the ladder builders were vying with one another to produce the best turntable ladder.

Although the turntable ladder originated in horse-drawn days it was the introduction of heavier motor chassis which allowed the ladder builder his head and provided a firmer foundaton for his design flair in producing better ladders.

As with the extension ladder the first turntable ladders were of mainly wooden construction with lengths gradually increasing as design and construction techniques improved.

Advertised as the all-British motor turntable ladder was the Merryweather which boasted ladder operation by propelling engine as far back as 1908. Typical of their production in the early 1930s is the vehicle shown in plate 93 which is a 100-ft all-steel ladder turned out from the Greenwich works for shipment to Hong Kong in 1933.

Leyland Fire Engines

The first Leyland fire engine was delivered to Dublin in 1910. This vehicle incorporated a 50-h.p. engine and a 250-gals-per-min. turbine pump and marked the beginning of a line of first-class fire appliances turned out during the following thirty years.

The early types of Leyland appliance were soon accepted by the brigades of the country and in the next few years a complete range of appliances was offered. The smallest was a four-cylinder 32-h.p. tender and the largest, a six-cylinder 85-h.p. pump with an output of up to 850 g.p.m.

During the 1920s the FE series was introduced and it continued until 1931 when the FT1 and TLM types appeared. 1933 saw the FK1 type added to the series and during the next few years very many variations were introduced to the FT and FK range.

The parts of many vehicles in the Leyland range of goods vehicles were interchangeable and so it was an easy matter to produce a non-standard vehicle using standard parts such as axles, engines and gearboxes should the need arise. Although the FK and FT range were based mainly on Cub and Tiger types, the Lioness, Terrier and Beaver were also used to a degree on a few occasions.

The FK type featured an 11 ft 6 in. wheelbase chassis with normal control, a six-cylinder petrol engine producing 62 b.h.p. driving through a four-speed gearbox to an overhead worm axle. The FT series was a larger vehicle with 12 ft 3 in. wheelbase and 114 b.h.p. engine. Both types could be supplied with various body styles and equipment to suit the particular brigade, with pumps of 400, 500 and 700 gallon capacities. There were many variations built during the life of the FK and FT range, involving pumps, wheelbases, engine type, forward control, and limousine enclosed bodywork.

Fig. I Leyland-Metz 100-ft turntable ladder (UK)

The TLM chassis was designed specifically for carrying the famous Metz turntable ladder and was of 14 ft 6 in. wheelbase with a six-cylinder o.h.v. petrol engine of 115 b.h.p. The Metz ladder was supplied in various lengths from 85 ft to 164 ft but the majority were of 85, 90 or 100 ft, with one particular machine going to Kingston-upon-Hull in 1935 with a ladder of 150 ft. A typical Leyland-Metz turntable ladder of the late 1930s is shown in fig. I.

Dennis Fire Engines *Plates 94–8*

The first Dennis fire engine was supplied to Bradford in 1908 for £900. Seven more were built in the same year, while the output for 1910 was twenty-seven, which included machines for Birmingham, London, and Glasgow. A total of 139 were built in 1929, and by 1934 over 2,000 machines had been supplied to brigades in many parts of the world to a total value of £1,691,168.

In the latter half of the 1920s the use of pneumatic tyres for heavier vehicles started to make headway, and in 1926 a Dennis pump of the Fort Dunlop fire brigade was so equipped. 1928 saw the introduction of the G type low-loading appliance as a 250/300-gal. machine, while in 1929 the first of a new type of 300/400-gal. appliance was delivered to the brigade in Birmingham. In the same year another innovation at Birmingham was their first machine with 'inside type' bodywork which could carry sixteen men on the inside of the body, protected from traffic accidents. This was the New World style of body.

1933 saw the production of a new type of rescue appliance for West Ham brigade which resembled a single-deck bus and was described as combining the assets of mobile power station, ambulance, canteen van and rescue tender all in one vehicle!

This was the era of the enclosed or 'limousine' appliance and vehicles of this type were supplied on various chassis. Sheffield had an enclosed 350/450-gal. four-cylinder machine based on the Ace chassis, while Eastbourne preferred the larger six-cylinder 110-b.h.p. model with 650/800-gal. pump. London had started its enclosed fleet with Dennis machines but later changed to Albion/Merryweather and Leyland types.

Illustrated are five Dennis fire engines from the 1930s. The first (plate 94) is a special hose-layer supplied to the London Fire Brigade in 1936. This machine was based on the Dennis Lancet chassis and carried $1\frac{1}{2}$ miles of hose. It was designed so that two lines of hose could be payed out over the sloping tailboard as the vehicle was driven forward at 15 m.p.h. Hinged flaps were provided along each side of the vehicle so that firemen could superintend the free flow of hose.

Next (plate 95) is a Canteen Van supplied to the London brigade in 1935. This vehicle used the Dennis 45-cwt forward-control chassis and was equipped with boilers, ovens, sink and dresser for the dispensing of food and drink to weary firemen at large fires. This vehicle replaced the old Commer Canteen Van illustrated in plate 59.

The third London machine illustrated is the Dennis crane lorry shown in plate 96. This impressive vehicle was supplied in 1936 and was based on the maker's 10 ft 6 in. wheelbase Lancet chassis with six-cylinder 120-b.h.p. engine. Equipment carried included an 8-ton hand-operated crane, oxy-acetylene cutting equipment, two- and four-wheel vehicle ambulances and heavy lifting jacks, blocks, crowbars, etc. When the vehicle was replaced in 1960 the Herbert Morris crane was transferred to the new chassis, which was also a Dennis.

As general fire appliances used by many brigades both at home and abroad, the 'Big 4' and 'Big 6' Dennis machines were both useful and impressive. Shown in plate 97 is a Big 4 pump which embodied the maker's 90-b.h.p. four-cylinder o.h.v. petrol engine driving through a four-speed gearbox to a worm axle. Bodywork of the machine was of ash framing with metal panels and in addition to the style shown, bodies of New World, double transverse seat and limousine pattern were available. A rear-mounted two-stage turbine pump was fitted, and this was rated at 800 gals-per-min. at 70 lbs per sq. in. pressure.

Of similar layout and style is the Big 6 pump escape shown in plate 98. This machine was fitted with the 110 × 140-mm. six-cylinder petrol engine which developed 115 b.h.p. and drove through a four-speed gearbox to a worm rear axle. The two-stage turbine pump was fitted at the rear although this could be positioned amidships if specified at time of ordering. A sliding carriage escape is carried on this machine, but extension ladders could be carried as required. As with the smaller four-cylinder machine this vehicle could be supplied with alternative bodywork if required.

Scammell Fire Engine   *Plate 99*
Three-wheeled fire engines are a comparative rarity and have not appeared in great numbers at any time.

During the 1930s a few were used by European brigades but their use was not widespread and with the introduction of standardised appliances they were gradually replaced.

In 1930 Meyer-Hagen offered a small appliance with one wheel at the rear and two at the front supporting an open box which carried a demountable pumping unit and some lengths of hose. This one-man appliance was based on the Goliath light truck.

A similar appliance was delivered to the brigade at Karlsruhe in 1932 but this used a Tempo type T10 three-wheeler with a two-stroke 10-h.p. engine. Inside the small-sided body a demountable pumping unit of 500-litre capacity and two hose reels were placed. There was a pillion seat behind the driver and two other seats were placed facing outwards immediately in front of the driver. How the driver ever saw his way to the fire with two burly firemen in front of him is anybody's guess!

In Great Britain just before World War II a few three-wheel appliances were built by County Commercial Cars. These had the single steering wheel at the front and the two wheels at the rear. Separate engines by Ford and Austin were used to provide motive power and to drive the rear-mounted pump.

A few appliances based on the Scammell Mechanical Horse chassis were also produced for wartime use and later in industrial premises. With seats for a crew of four, extension ladder, 350-gal. water tank, twin hose reels and a rear-mounted Scammell pump this appliance looks a little incongruous, but was nevertheless useful (plate 99).

Earlier in the 1930s Scammell had built two full size fire engines for use by the local brigade in Watford, Hertfordshire. The first of these was delivered in 1933 and the second in 1935.

Ford Fire Engines *Plates 100-1*

From the early days of the Model T Ford chassis have formed the basis for many fire engines in service in various parts of the world.

The introduction of the V8 engine in 1935 marked another step forward in the use of Ford vehicles to provide high speed appliances at moderate cost, and many fire engine builders turned out machines with the V8 power unit including the Rumanian builder whose machine is illustrated in plate 100.

The Ford V8 appliance shown in plate 101 is a type specially built to the requirements of the British Home Office in 1939 by the Ford Motor Company.

During the few years before World War II the Home Office had been issuing design specifications for pumping units, trailer pumps, emergency fire appliances and emergency water dams. Various manufacturers were supplying this equipment as the threat of war increased and the popular Ford 7V chassis was used for mounting heavy pumping units and for carrying escapes.

The appliance shown in plate 101 is a 2-ton chassis with 9 ft 10 in. wheelbase and the well-known Ford V8 petrol engine. The cab provided accommodation for the crew of six and had a specially strengthened double-skinned roof to protect the crew from falling debris. A roof gantry carried extension ladders, side lockers were provided for rolled hose storage and the suction hose was strapped to the vehicle body. The Sulzer two-stage centrifugal pump, with an output of 700 gallons per minute, and its Ford V8 power unit were mounted on a separate frame for ease of production and removal should the chassis become useless.

Emergency Towing Vehicles *Plate 102*

With most of the commercial vehicle production going to the armed forces during the period of the 1939–45 war, only a limited number of small- to medium-capacity vehicles were available for use by the fire and civil defence authorities. Before the war some brigades had fitted their regular appliances with towing equipment for the trailer pumps distributed by the Home Office and brigades such as that in London had bought Ford 15-cwt vans specially for the purpose of towing trailer pumps.

In order that many of the trailer pumps could be stationed at strategic points remote from the regular fire stations it was necessary to introduce other towing vehicles to keep the fire columns mobile. This resulted in many large cars being fitted with towing gear and other fitments for use as emergency fire-fighting units. In addition quite a number of London-type taxicabs were similarly fitted in order to render useful service. Many larger private cars with long wheelbase also had the body removed from a point just behind the front doors, and ambulance bodies fitted.

The vehicle shown in plate 102 is a Ford V8 saloon car chassis with the rear of the body removed and a utility van body fabricated to carry the fire crew on two longitudinal seats. Two ladders were carried on the roof and small items of equipment were stowed within the van body.

Trailer Pumps *Plates 103–4*

The modern concept of a trailer pump is a two-wheeled self-contained pumping unit complete with its own power source. The trailer pump is one of the oldest ideas of a pump and in the past there have been pumps trailed by hand, horse and motor power.

The early manual and steam pumps were trailers and there have been a few motor pumps trailed by horses (plate 34). In some instances pumps originally designed to be pulled by horses were later modified by removing the draught pole and substituting a draw-bar for coupling to a hose and ladder truck or tender vehicle.

Trailer pumps as we know them today originated in the 1920s at about the same time as Tilling Stevens introduced their demountable pump which was in effect a pump with driving electric motor and mounted on two wheels. It received its electrical supply from the dynamo mounted on the parent vehicle and could be positioned as far away as the length of power cable would allow.

Dennis introduced their trailer pump in 1922 and the first Merryweather Hatfield went to Cairo. From that time the trailer pump gradually gained favour as a useful adjunct to fighting fires where accessibility for a full size motor pump is difficult.

So far as Britain was concerned it was World War II which put the trailer pump into its correct perspective because it was realised that the country couldn't hope to mobilise the great many motor pumps required to resist large-scale bombing raids.

In 1937 the British Home Office issued a memorandum on emergency fire brigade organisation and stated that the government would provide free apparatus of an approved type. The first approved types were self-contained pumping units for mounting on lorries but later on specifications were issued for trailer pumps of various sizes. Some of these were heavy pumping units capable of producing 700 gallons per minute at 100 lbs pressure and had to be mounted on four-wheel drawbar trailer chassis because of their weight.

From 1937 onwards lighter trailer pumps started to be produced and gradually the familiar names of fire appliance builders were joined by others less familiar as the Home Office scheme gained momentum. The pumps were divided into three categories, large, medium and light, and their output varied from 500 g.p.m down to 140 g.p.m.

Some of the trailer pumps soon gained the admiration of their crews and were kept in service after the war because of their constant reliability. Others were not so fortunate and at the denationalisation of the fire service in 1948 individual areas tried to keep the pumps that they liked.

One of the outstanding pumps of the period was the Coventry Climax, and it was this company under the leadership of Leonard P. Lee that produced over 25,000 pumps for the fire service and armed forces. Since those wartime days the pumps have been further improved and the original FF and FSM types have been replaced by a series of lightweight trailer and demountable pumps as well as vehicle-mounted types.

Supply and Support Vehicles *Plates 105–6*

Although the majority of vehicles in service with the fire brigades are of a type designed for actual fire fighting, there are others which have a functon as supply or support vehicles. Some large brigades employ many such vehicles in a variety of roles. Small vans may be used to distribute stores, supplies, uniforms and equipment. Lorries are used to carry hose, foam, tools, ladders, etc. and large vans have been adapted as control rooms, telephone exchanges, radio stations, and even as emergency fire stations.

In recent years two buses have been used by fire brigades as mobile control rooms for use at large fires. A Leyland double-decker was used for many years by Kent Fire Brigade for this purpose, and the London Fire Brigade obtained an ex-London Transport Leyland Cub one-and-a-half-decker for a similar function.

During the years following World War II the British Home Office set about providing mobile fire columns quite distinct from the normal brigades of the country. The intention was that these columns should be completely self-contained and be capable of moving to disaster areas at short notice. Naturally there were many fire-fighting vehicles within the columns, but there were also a number of support and supply vehicles needed to maintain the fire column in an area which may be totally disrupted by enemy action. These ancillary vehicles took the form of mobile control units, command cars, communications units, pipe carriers, boat carriers and mobile kitchens.

The importance of keeping active firemen well fed had not gone unnoticed, and many years earlier complete vehicles had been provided to maintain a supply of food and drink. (Plates 59 and 95.)

The first large-scale emergency to face the British firemen had been World War II and in order to keep the fire engine crews supplied with food during long nights of fire fighting a number of mobile kitchens were put into service in large towns. These kitchens were based mainly on Ford 7V chassis (plate 105) and contained cooker, cupboards, boilers, racks and the necessary crockery.

For the supply of that unique soul mender, the British cup of tea, many small canteen vans were put into service for use by rescue, civil defence, police and firemen. Many of these were mounted on Ford 10-cwt van chassis, others were supplied from the Dominions overseas and used American or Canadian light van chassis. As mentioned elsewhere, the country was short of all kinds of vehicle so many large private cars had the bodies removed and a spartan type of canteen body imposed. The Humber canteen in plate 106 shows a typical conversion.

Carl Metz *Plate 128*

Mention the name Metz to anyone remotely connected with fire fighting and he will immediately think of turntable ladders, but this is not the only production of this world famous name.

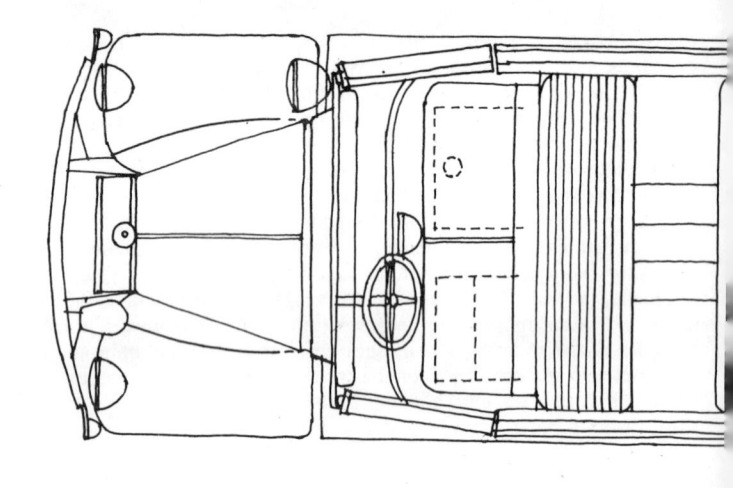

Fig. J Mercedes Benz-Metz LF 15, 1947
(Germany)

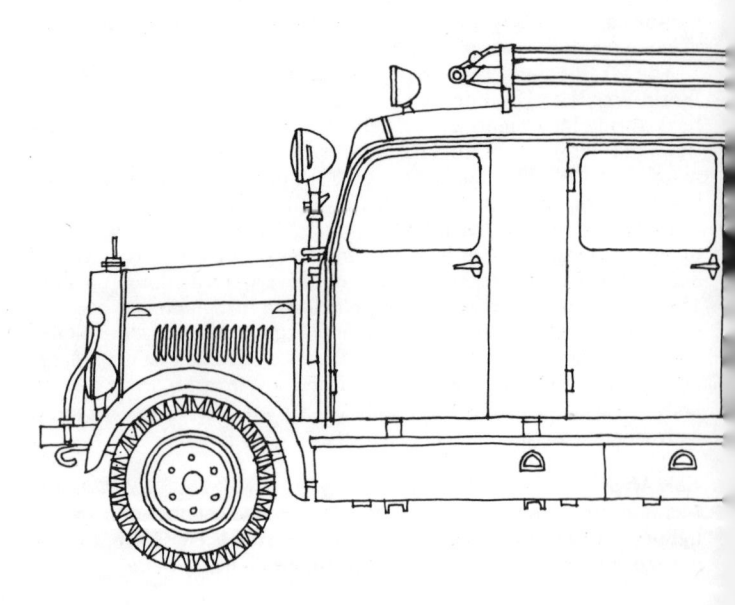

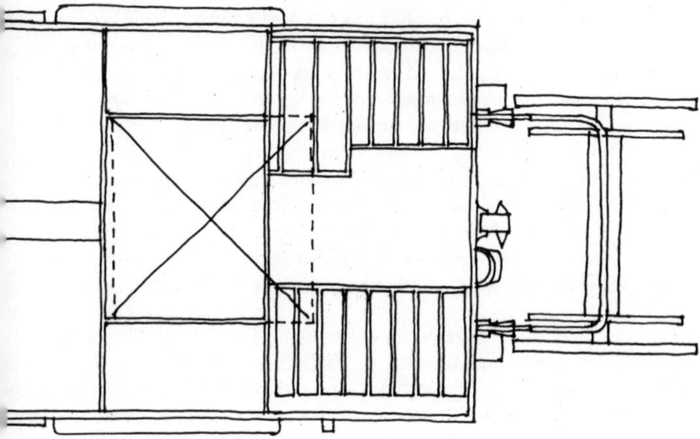

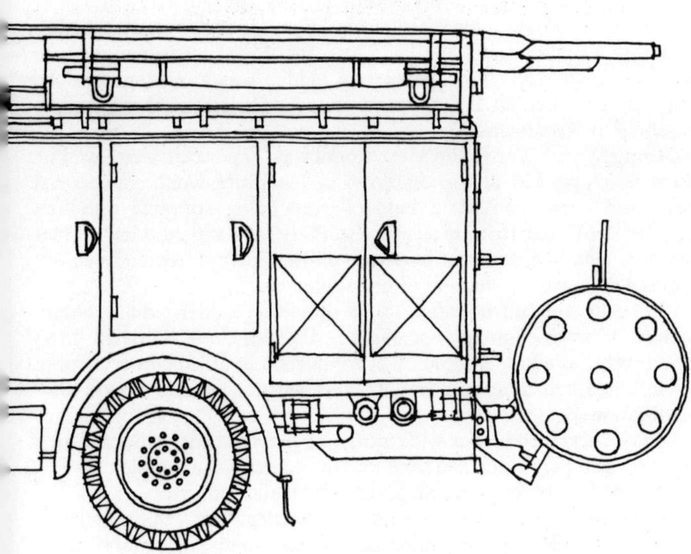

It was in 1842 that the mechanic Carl Metz founded his works at Heidelberg and it remained there until 1904. In 1846 he founded the local volunteer fire brigade, and in 1893 his company was headed by a board of directors. Metz lost control of the company to Alfred and Karl Bachert in 1905 and they moved the works to Karlsruhe.

From 1906 the company turned its attention to petrol-engined fire engines instead of the manual, horse-drawn and steam ones built previously. They used the vehicle-building knowledge of Suddeutsche Automobilfabrik of Gaggenau/Baden to carry their fire fighting equipment and this close co-operation has continued with the Daimler Benz concern which succeeded the Suddeutsche concern in 1908.

1912 saw the first of the now famous Metz turntable ladders delivered to the local fire brigade of Karlsruhe, by 1918 the ladders incorporated mechanical safety devices and in 1924 the first Metz all-metal ladder appeared.

In 1933 Metz built a fire appliance on a diesel-engined chassis and in 1935 obtained a patent for turntable ladders with hydraulic gear. Although there was competition from other ladder builders in Europe the Metz ladder was extremely popular in many countries, and in Great Britain an agreement was signed with Leyland Motors for Metz ladders to be mounted on Leyland chassis for all sales in Great Britain and its Commonwealth. Many 30-m. ladders were supplied to towns in England and in 1935 one of 45 m. was supplied to Kingston-upon-Hull. This particular ladder was the longest to be supplied to a British brigade and this significant machine is now preserved by Tim Nicholson.

During World War II the Metz works was 70 per cent destroyed by Allied bombing but at the cessation of hostilities work commenced again and very soon Metz ladders were being supplied to many brigades. In Great Britain Dennis Brothers took up the Metz ladder franchise which was vacant by virtue of the fact that Leyland had not resumed post-war production of fire appliances.

1957 saw the introduction of a 60-metre ladder and a ladder capable of cross-country movement – the DL 25 – while in 1958 Metz produced a ladder which was completely hydraulic in operation. Ladders adapted as cranes and ladders with cages and rescue platforms followed.

Today Metz produce a wide range of fire-fighting vehicles which includes pumps, water tenders, rescue vehicles, turntable ladders, airport crash tenders, portable pumps and trailer pumps.

Shown in fig. J is a Metz-equipped Mercedes Benz vehicle of 1947. This is the LF 15 type, mounted on a 3-ton chassis, incorporating an

80-h.p. four-cylinder diesel engine. A rear-mounted centrifugal pump delivers 1,500 litres a minute and is fed from a hydrant, or from open water, or from the inboard 400-litre water tank. Behind the driver is the crew cab, and lockers for rolled hose, lamps, tools, couplings and fittings are on both sides of the body and above the pump at the rear. The roof has gantries for extending ladders and a stretcher while a demountable hose reel or devidor is mounted at the extreme end of the appliance.

The layout of this machine is typical of the many hundreds which were supplied to the German fire service during the 1930s and 1940s after it was announced that a standardisation programme was to be instituted. This standardisation covered mobile appliances, hose, branch-pipes, couplings and hydrants, and was aimed at promoting an efficient fire service with standard and interchangeable equipment. Another startling announcement made at the 1935 International Fire Prevention and Public Security Committee exhibition in Dresden was that in future all German fire apparatus would be diesel-engined.

Although Metz equipment was usually mounted on Daimler Benz chassis, in the period up to 1940 other marques to carry their products included DAAG, NAG, Bussing, Opel, Hansa-Lloyd, Adler, Durkopp, Leyland, MAN and Faun. Following World War II Metz fire equipment has appeared on a variety of chassis by such makers as Saviem, Bussing, Citroen, DAF, Dennis, Ford, Faun, Fiat, Kaelble, MAN, Steyr, Saurer, Scania, Tatra, Unic, Volvo, ERF, FBW, Leyland, Berliet, Bedford and Commer.

Mobile Control Unit *Plate 107*

With the majority of fires only a handful of appliances are needed and the senior officer present at the fireground can control the men and equipment at his disposal from his position there. Should he decide that additional men and equipment are required then he is at liberty to summon them should he so desire. Naturally the position and classification of the fire risk will determine the number and type of appliances that will be summoned, and the status of the officer-in-charge will be commensurate with the size of fire and risks involved.

Should the situation, because of its size or duration or complexity, demand special attention it is likely that a control post will be set up adjacent to the fireground in order that the whole situation can be better assessed and controlled. This state of affairs is likely to prevail in war time when the whole position of fighting the fire, maintaining supplies and ensuring contact is in jeopardy because of enemy action. In order to combat this, Mobile Control Units were designed to assist the brigade in its task of maintaining adequate fire-fighting facilities in the face of all the disasters which may occur.

The vehicle illustrated in plate 107 is a Mobile Control Unit mounted on a Ford 7V chassis with V8 petrol engine. Basically a mobile office and communications post, it possessed facilities for the centralising of command at a large fire or was used where a fire station had been destroyed by enemy action.

Bedford-Pyrene Airfield Crash Tender   *Plate 109*
The pre-requisites of an airfield crash tender are briefly (1) that the machine should be capable of being started immediately and driven to the scene of the crash at high speed, (2) that the appliance should be completely self-contained so that upon arrival at the crash the equipment necessary is available at once and (3) that the appliance should produce the largest amount of foam possible extremely swiftly and that the skilled operators should be able to spread it effectively.

With these salient points in mind the designer of such a machine sets about producing a vehicle which will meet the demands as far as possible. For many years no special vehicle chassis were available on which to mount the heavy body and fire-fighting equipment needed nor which were capable of the rapid acceleration demanded for the short high-speed run across an airfield. Various types of crash tender were in use but most of these were mounted on the available proprietary chassis. In fact some of them were extremely stark in appearance consisting of nothing more than a foam-producing unit mounted on a lorry chassis/cab.

With the coming of the European war the need for something more specific became apparent and for airfield duties in Britain and overseas appliances based on Crossley, Austin and Ford chassis were put into service.

One of the most outstanding vehicle chassis produced during the war was the Bedford QL four-wheel-drive 3-tonner. This chassis was used as a basis for many types of body for HM Forces and it is not surprising that at the end of hostilities the same chassis was specified to carry the body, pump and equipment marketed as the Pyrene Airfield Crash Tender (plate 109).

Brief details of the machine are wheelbase 11 ft 11 in., track, front 5 ft 8 in., rear 5 ft $6\frac{1}{2}$ in., turning circle 54 ft; both axles driven, spiral bevel gearing, fully floating axle shafts; Lockheed hydraulic servo-assisted brakes, 15-gal. fuel tank common to both road and pump-engines; six-cylinder Bedford SV petrol engine $3\frac{3}{8} \times 4$ in. producing 72 b.h.p. at 3,000 r.p.m. driving through a 10-in.-diameter single dry-plate clutch and a four-speed gearbox to a centrally-mounted transfer box which transmits the drive to front and rear axles. Either two- or four-wheel drive could be selected. A separate Coventry-Climax F-type engine was used to power the FF-type 500-gals-per-min. single-stage centrifugal pump of the same manufacture. This engine was a four-cylinder side-valve unit producing 75 b.h.p. at 3,000 r.p.m. from its four 90 × 130-mm. cylinders.

The equipment of the machine was disposed as follows: in the two lockers either side of the radiator grille at the front of the vehicle were stowed two 80-ft lengths of 4-in. canvas hose and nozzles. Behind these two lockers the two foam generators were mounted shrouded into the front wings. Foam was produced by the pump taking water from the 500-gal. tank mounted longitudinally over the rear axle, feeding it direct to the two foam generators where the foam compound was added. The foam compound was stored in a 40-gal. tank mounted on the nearside just in front of the rear axle. The enclosed cab accommodation was in three bays – the first for driver and officer, the second for the transversely-mounted pump and the third for the remainder of the crew and battery of six 50-lb. CO_2 gas cylinders. Two large lockers on either side of the water tank housed the CO_2 hose reels and gas distributors. The roof of the water tank and the two side catwalks were designed for storing suction hoses and ladders as required.

Pyrene Fire Equipment *Plates 110–18*

Although never having built any complete fire appliances themselves, the Pyrene concern have supplied equipment for a wide variety of applications. In plates 110 to 118 are shown some of the productions ranging from high-output oilfield tenders down to the diminutive Vespa scooter and even wheeled fire extinguishers.

We all imagine that airfield fires are caused by aircraft catching fire upon crashing or landing, but the Vespa lightweight combination fire-fighting unit shown in plate 110 is a unit designed to extinguish small fires generated by aircraft when starting up. It is equipped with two 12-lb CO_2 cylinders which feed through gas-release valves and high-pressure hose to an extending applicator which enables the CO_2 gas to be discharged direct into the engine nacelles.

A type of appliance utilizing an ordinary small truck loaded with cylinders of liquid carbon dioxide and foam cylinders is the Morris Commercial 15-cwt of wartime fame. This sort of apparatus would be of value in tackling small outbreaks in industrial premises (plate 111).

A more usual type of CO_2 tender is shown in plate 112. This uses a Dodge chassis and is designed to dispense quickly a large volume of

CO_2 gas for extinguishing fires fed by alcohols, solvents, naphtha, oil or spirits. The use of CO_2 gas as an extinguishing agent is also of value when fighting fires in electrical equipment such as switchgear and transformers. The vehicle carries the liquid CO_2 in high-pressure cylinders, and the discharge nozzles of which are manifolded in banks and fed to rear-mounted hose reels and hooded applicators. A supply of foam compound is carried in 2-gal. cans and these are used in conjunction with the water tank mounted over the rear axle to produce expanded foam for oil fires.

Another type of CO_2 and foam tender is shown in plate 113. This uses a Bedford six-wheel chassis and is equipped with a large foam monitor at the rear.

Plate 114 shows another style of foam monitor mounted on a Bedford 4×4 chassis. A 500-gal. foam compound tank feeds direct to the monitor or to smaller hand lines at the rear and the vehicle can be replenished by connection to a foam tanker supply vehicle.

A larger CO_2 and foam tender based on a 6×6 Scammell Constructor chassis is shown in plate 115. This machine is equipped with a large foam compound tank and banks of liquid CO_2 cylinders. A large-capacity monitor is used for fighting fires from a distance while connections for normal hose are provided at the rear and both sides. Two self-contained engine/pump units are mounted at the rear and can be used independently or combined to produce 1,800 gals. of foam per minute.

To be able to fight a large fire from many different points is a great advantage, but with the intense heat generated by some fires it is necessary to keep a safe distance. A large volume of the extinguishant is also required but if the large-capacity monitor is fixed to the vehicle then only one position is possible. In an attempt to overcome this problem ground monitors are often used, but with the large volume of foam required in oil fires the sheer bulk of the monitor makes manual handling impracticable. The trailer mounted monitor shown in plate 116 is a solution to the problem in that it is quickly transported to the scene, easily moved to a new position and readily connected to other mobile appliances or fixed foam mains.

The Bedford water tanker shown in plate 117 is equipped with a 1,000-gal. tank and a rear-mounted Coventry-Climax pump. Although looking rather spartan in appearance compared with the pump/water tenders normally used by fire brigades, the vehicle is of great use where water supply is either restricted or remote. In industrial applications it can quickly provide a large amount of water to 'knock down' a fire in its early stages or it may be used purely to supply other pumping appliances.

Alvis Airfield Crash Tender
Plate 119

One of the most unusual modern fire-fighting units is the Alvis Salamander six-wheeled Airfield Crash Tender as shown in plate 119. This machine is really the civilian version of the Stalwart military vehicle famous for its amphibious capabilities. In fact two of the Stalwart vehicles have been produced for use as emergency vehicles, one for use as a rescue vehicle and crash tender for Rangoon Harbour and a second for use by the Royal Air Force at Guan Island in the Pacific, where it is used as an airfield and amphibious rescue vehicle.

Salamander fire appliances are produced with specialist bodywork and equipment by either Pyrene or Foamite and the vehicle illustrated is one equipped by Pyrene.

The complete vehicle measures 18 ft long, 8 ft 3 in. wide and 10 ft high, weighs 12 tons 8 cwt and has a turning circle of 45 ft. As well as possessing drive on all wheels the cross-country ability of the vehicle is further enhanced by virtue of its ground clearance of 1 ft 6 in. and clearance angles of 45° both on approach and departure. The first two axles steer. Power is supplied by a Rolls Royce B81 six-cylinder petrol engine producing 240 b.h.p. which is positioned at the rear of the vehicle.

Land-Rover
Plates 120-1

There has always been a need for small fire-fighting apparatus and in the early days this need was met by hand-drawn or trailer appliances which were small and light enough to be manhandled into confined areas or along unmade roads. To a certain extent this state of affairs still prevails today, and although the amount of concrete covering the country is still increasing there are still areas remote from roads over which our fire-fighters have to travel. In another instance we have the increasing complexity of modern cities where there is a definite trend toward the separation of traffic roads from areas reserved for living, shopping or recreation. Often it is areas such as these that rate high in the fire risk areas of towns.

From the needs of the two circumstances mentioned there has been derived one machine with certain characteristics, although on the face of it the two risks are dissimilar. This has been a vehicle of small overall size, highly mobile, of exceptional climbing ability, reasonably fast, and light in weight, carrying men and equipment based on first aid principle.

In Great Britain and in many other countries this need has been filled by the Land-Rover produced by the Rover Company, now a division of the British Leyland Motor Corporation. Many vehicles of this type have been produced with bodies and equipment by the specialist suppliers of fire apparatus.

The vehicle illustrated in plate 121 shows a standard 109-in.-wheelbase Land-Rover with specialist bodywork and fittings by HCB-Angus. This model is termed the Firefly.

Although the Land-Rover has been the answer to many problems for a compact, go-anywhere vehicle capable of 101 uses one drawback has been the fact that the body space has been small. With all normal-control chassis a lot of the overall length is taken up with the bonnet and cab, so faced with the dilemma of crowded body space Carmichael & Sons produced a forward-control conversion which gave extra space for crew and equipment (plate 120). The Redwing FT6 model offered seating for a crew of four, a 100-gal. water tank, choice of pumps in the range of 300 to 500 g.p.m., and considerable locker space for hose.

Maxim *Plates 122–4*

The building of fire apparatus by Maxim goes back to 1914 when they produced a combination fire truck for the Middlesboro Fire Department. Orders for similar machines for other New England area towns followed, and it soon became evident that designs should be prepared for pumpers and ladder trucks to provide a more complete line of fire apparatus.

In 1918 the business was incorporated under the name Maxim Motor Company and additional capital was acquired from investors in addition to that held by the Maxim family. At about this time the company started to build commercial trucks but this project did not last long. Over the years the company continued to prosper and the premises were enlarged.

In common with many other companies Maxim was hard pressed during the period 1929 to 1939 because of the financial crisis following the stock market crash. The fire truck market was highly competitive and Maxim ventured back into the commercial truck field for a short while.

The Second World War saw the company building fire trucks, airfield crash tenders and other equipment for the US Government, and although these were curtailed at the end of hostilities Maxim launched a new series of fire trucks which had been designed ready for the return of peace-time production. The company was also able to go ahead with the production of a design of aerial ladder which they had been contemplating many years earlier.

In 1956 control of the company passed from the Maxim family to that of the Seagrave Corporation.

Today a line of fire trucks is produced under the Maxim name ranging from the 127-in.-long 2800 Mini-Max truck, through pumpers, ladder trucks, booster tanks, rescue trucks and brush trucks to articulated aerials and airfield crash tenders.

At the lower end of the Maxim range is a vehicle which must surely qualify as the most diminutive four-wheeled fire appliance anywhere in the world. Aptly named the Mini-Max, this novel piece of equipment meets the need for a modest little fire-fighter to protect factories, works, warehouses, colleges and stadia. The tiny vehicle illustrated in plate 122 is equipped with a 150-lb. dry chemical unit, a 20-lb. dry powder extinguisher, a $2\frac{1}{2}$-gal. water pressure extinguisher, a two-section 12-ft extension ladder, two fire axes and a back-pack type of breathing apparatus. In addition there are fire lockers for additional equipment and a two-tone horn and flashing lights. There are seats at the front for two crew members while a small rear platform can accommodate others. The vehicle measures 127 in. long, 47 in. wide and 67 in. high and weighs 2,800 lbs. It is available with alternative body styles and with power by petrol, propane or batteries. Another way in which this gallant little vehicle is unique is the fact that it is also offered with optional remote control!

The range of Maxim aerial ladders is exemplified by the apparatus shown in plate 123 which is a front-mounted 100-ft four-section ladder built on an articulated vehicle. Designated FF-CLT, this outfit has the closed crew cab but an open style cab is available if required (model designation FF-OLT). Also available is an articulated aerial ladder of up to 100 ft but of the engine-ahead design, designated SO-LT or SC-LT according to whether an open or a closed cab is used.

A range of ladders mounted on four-wheel chassis is produced and these are available with normal or forward control, open or closed cab styles, and front- or rear-mounted ladders. The ladders are all hydraulically operated, with the operating console positioned alongside the ladder base on the rotating platform. An optional extra is the provision of a duplicate set of controls for the fireman at the top of the ladder with overriding control still in the hands of the operator at the base of the ladder in case of any emergency.

Illustrated in plate 124 is a 4×4 Brush Truck equipped with a main pump, portable pumps, water tank, hose reels, flaked hose and full brush guards at front, sides and overhead. The International chassis is specially protected for travelling through blazing brush by means of underpanning.

One of the company's earlier apparatuses is shown in fig. K. This is a long-wheelbase four-wheel hook, ladder and chemical wagon of the 1920s.

Leyland Firemaster Appliance   *Plate 125*

Leyland Motors have produced fire appliances on a variety of chassis from the time when they entered the fire engine market in 1910 by supplying a complete vehicle to Dublin. They are generally acknowledged to have produced some of the most useful and attractive machines for use in Great Britain and abroad, and this company was at the zenith of its history so far as fire-fighting vehicles are concerned in the years leading up to World War II.

Many people expected that this famous marque would reappear soon after 1945, but it was not to be, for they confined their production to commercial vehicles, buses and coaches, although a brief entry (if one can call it that) was made a few years after the war when the bonneted version of the Comet goods chassis was used as a basis for a small number of fire appliances.

However, things took on a different light in 1958 when the Leyland Firemaster made its appearance, and many thought that this was to be the progenitor of a new Leyland breed of vehicles for fire-fighting. But they were mistaken. During the ensuing five years only a handful of these appliances were built, so what could have been the re-entry of Britain's largest vehicle producer into fire appliances did not materialize. Strictly speaking the Firemaster was not a Leyland fire engine in the same sense as the pre-war range, because as with so many other fire engines it was the chassis only that was produced by Leyland, unlike the previous productions which mostly

Fig. K Maxim chemical hook and ladder (USA)

had bodywork built in the company's shops. It is true that from time to time other Leyland chassis have been used as a basis for fire engines and continue to be but none, other than the Leyland Albion Firechief, has been advertised as such.

The basis of the Firemaster was a straight coach-type chassis embodying features of Leyland coach practice at the time. A Leyland 0.600 six-cylinder CI engine with a capacity of 9·8 litres powered the vehicle and produced 150 b.h.p. at 2,200 r.p.m. Drive was by means of a Leyland semi-automatic transmission with two-pedal control. The engine was mounted amidships and like the other mechanical units was below chassis frame level, thus providing the bodybuilder with an uncluttered floor line. On those machines that carried pumps they were placed at the front but within the lines of the body, not right out in front as with most other front-mounted pump appliances.

Although only ten vehicles were built they were certainly a mixed bunch so far as fire appliance types are concerned, there being pump-escape, turntable ladder, emergency tender and Pump-Emergency/Salvage Tenders built during the period up to 1963 when the last was delivered to Manchester Fire Brigade.

Dennis F Range   *Plates 126–8*

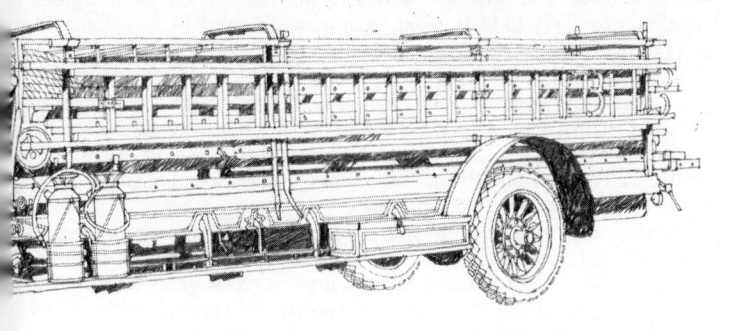

Soon after World War II motor manufacturers were beginning to resume peace-time production with vehicles to replace those that had been in service for so long but because of the needs of war had been kept running. The Dennis range of fire engines began with the F1 or Onslow type which was based on a bonneted chassis with four-cylinder 70 b.h.p. petrol engine. The rear-mounted pump was of a multi-stage turbine type producing 400 gals per min. at a pressure of 120 lbs per sq. in., and the body could be of either Braidwood or New World pattern as required. This machine was intended for rural brigades and was only seven feet wide, very useful for negotiating narrow country roads.

For a machine with a better road performance the F2 appliance was available with a Rolls Royce B80 straight eight-cylinder petrol engine. Whilst still retaining a narrow width, the wheelbase of this machine was 13 ft 6 in. compared with the 10 ft 6 in. of the F1, thereby providing additional body space. The machine illustrated in plate 126 is a Pyrene foam tender designed for oilfield fire protection. This vehicle has the 900 g.p.m. pump mounted amidships with control and hose connections on either side. A 600-gal. foam compound tank is carried, and when the foam compound is induced at the water heads of the mechanical foam generators up to 120,000 gal. of foam are produced.

With the exception of vehicles specially prepared, the last type of Dennis fire engine to incorporate normal control was the F6 tanker type appliance. This machine was based on the current municipal vehicle chassis with 11 ft 6 in. wheelbase, 100 × 120-mm. side-valve petrol engine, and 34 × 7-in. tyres. A 600-gal. cylindrical water tank, rear mounted pump and hose reel were supplied, together with overhead gantry for extension ladders. An optional extra was a set of sprinkler heads for street washing!

The more widely used forward-control range of fire engines began with the F7 F12 F14 trio which was an entirely new design of chassis with wheelbases of 13 ft 6 in. for the F12 and F14, and 12 ft. 6 in. for the F7. The F7 and F12 were designed as pumping machines to carry ladders or an escape, while the F14 was designed to take the Metz 100-ft turntable ladder (plate 128). Dennis had been successful in negotiating the franchise of this piece of equipment for their post-war range because Leylands, who had held the franchise for the British Empire since 1911, had decided not to re-enter the fire engine market after the war. All these three types of appliance incorporated the Rolls Royce B80 eight-cylinder petrol engine which gave the vehicles a lively performance.

As the new range of appliances was gradually improved over the next few years so the design of the front end underwent various changes. The F12 radiator grille was a plain rectangular opening in the sheet metal front end, but the shorter F7 sported an exterior chrome radiator shell in the established Dennis style. The Dennis pump escape in Coventry yellow colour scheme shown in plate 127 is of this type.

Today the Dennis range is made up of four basic types of appliance.

The D series, which is the smallest in the range, is based on a compact chassis of 10 ft 9 in. wheelbase and only 7 ft overall width. Power units are Jaguar 4·2-litre six-cylinder petrol or Perkins T6.354 six-cylinder diesel engine, and transmission is by way of Allison four-speed full automatic or Turner five-speed manual gearbox. A variety of bodywork is available which includes water tender, pump escape, emergency tender and special purpose tender applications.

The well-proven F series is based on a chassis having great similarity with the D series save that it can be obtained in two wheelbases – 12 ft 6 in. and 14 ft 3 in. according to application – and is 7 ft 6 in. wide.

A Rolls Royce petrol engine or a Perkins V8 diesel engine can be specified and automatic or manual gearboxes are offered. A choice of bodywork is available covering water tender, water tender ladder, pump escape, 50-ft snorkel, emergency tender or special purpose tenders.

For the mounting of turntable ladders and greater length snorkels a Special Purpose Appliance chassis is produced with a wheelbase of 16 ft 3 in and 8 ft $2\frac{1}{2}$ in. overall width. Power units and gearboxes offered are as in the F series but with this chassis a special low-profile cab is provided which puts the driver ahead of the engine and obtains a low overall height. This chassis is designed for the mounting of 70-ft or 85-ft snorkels and 100-ft or 125-ft turntable ladders.

The range of airfield crash tenders is of three distinct types, the 4 × 4 Standard with power supplied by either Rolls Royce B8I petrol or Perkins V8 diesel engine; the 6 × 6 Major based on the Thornycroft Nubian Major chassis with Cummins V8-300 diesel engine; and the 6 × 6 Ultra, which is a machine designed to meet the arduous duties of large airfields and is powered by a Cummins VTA 1710-700 diesel engine developing 700 b.h.p. at 2,100 r.p.m.

AEC-Merryweather *Plate 129*

Although the firm of AEC is an old-established one in the commercial vehicle market they did not produce any vehicles for fire-fighting until after the 1939–45 war. However, AEC did not take up the opportunities available in the need for new fire engines to replace those pre-war machines which had worked so hard, it was left to Merryweather to utilise AEC chassis when they were looking for a proprietary commercial chassis upon which to mount their equipment. The chassis chosen was the Regent Mark III, the basic design of which had been well proven over the years by its extensive use by bus operators in Great Britain and many countries overseas. The Regent chassis was designed for motor bus work and as such had a low chassis to accommodate double-deck bodywork. This feature is useful for fire-appliance work because it provides the vehicle with a lower centre of gravity for a given height compared with a similar machine mounted on a lorry-type chassis.

The machine illustrated in plate 129 is a Merryweather Marquis – the registered name adopted by Merryweather for their post-war series of high performance fire engines mounted on a Maudslay or AEC chassis. This particular machine is based on an AEC Mercury chassis and dates from 1963. It is basically a normal goods vehicle chassis suitably modified regarding power take off for pump drive and was supplied to Merryweather in chassis/scuttle form. The AEC AV470 six-cylinder diesel engine producing 130 b.h.p. was the power unit driving a five-speed gearbox. The Merryweather bodywork provided a crew cab with sliding doors on either side, locker space behind the bulkhead and low down at the sides. A single-stage turbine pump was offered as either midships or rear mounting. This pump was rated at 1,000 gals per min. with pressure ranging up to 250 lbs per sq. in. when used with the small-bore hose reel. Water tank capacities of 100 to 500 gals were available, and the tank was mounted above the rear axle. Hose reels were generally mounted either side of the water tank. If required a large space could be provided low down on the nearside of the machine for the storage of a demountable pump.

Japanese Fire Engines *Plates 130–2*

The largest vehicle in the Hino range of chassis for fire-fighting purposes is the Hino TC321 twin steering six wheel 32 m. turntable ladder (plate 130 and fig. L). Power for the vehicle is by means of a Hino DK10 six-cylinder diesel engine of 10,178 c.c. capacity which produces 200 h.p. at 2,300 r.p.m. Drive is taken through a six-speed manual gearbox to a hypoid gear rear axle. Wheelbase is 6,300 mm. and the vehicle is carried on $10 \cdot 00 \times 20$ tyres. A top speed of 95 k.p.h. is claimed for the vehicle, which has a total all-up weight of 16,645 kg. Braking is by means of an electro-pneumatic exhaust brake, full air brakes to all wheels, and a transmission parking brake. The all-steel cab has seats for three, including driver.

The five-section metal ladder extends to a vertical height of 32 m. and is equipped with a lift on the top side which is operated by means of a steel wire wound on and off a drum. Actuation of the ladder is by hydraulic means, power being supplied by the vehicle engine driving through a power take off. Extension of the ladder is by means of a winding drum and steel cables. The ladder is equipped with safety devices to prevent the accidental exceeding of any of the safety limits set by the manufacturers with regard to load, extension, inclination, falling, contraction or side movement. Four extending hydraulic jacks are positioned adjacent to the ladder base to ensure stability when the ladder is in the extended position, and in addition to these a rear-spring locking device is provided to prevent chassis distortion.

Pumping power is by means of a midships-positioned two-stage turbine pump with an output of 2,250 litres per minute at a pressure of 8·5 kg. per sq. cm. (500 gals at 120 lbs per sq. in.), which draws water from suction inlets mounted on either side of the body, and can discharge through two outlets on either side, or through a monitor mounted at the head of the ladder.

Water fog spray nozzles for the protection of the crew are fitted to both sides of the body as well as to the operating position at the head of the ladder. A total of four searchlights can be supplied to the vehicle, one at the head of the ladder, another at the ladder base, and two others located on top of the body just in front of the pump position. Additional equipment consists of suction hose, delivery hose, fire hook, axe, crowbar and ropes.

In contrast to the Hino aerial ladder are two smaller Japanese appliances, the Isuzu GD150F shown in plate 131 and the Toyota FJ45 shown in plate 132. Both these machines show a distinct American influence in that they have a midships-mounted pump, hose flaked flat in the rear body compartment, and crew positioned on the rear platform with a grab rail for security.

The Isuzu is powered by a six-cylinder 5,640-c.c. engine developing 145 b.h.p. and is capable of 85 k.p.h. with an all-up weight of 5,050 kg. With a wheelbase of 3,000 mm. it is 6,503 mm. long and 1,840 mm. high and carried on 700 × 20 tyres. The Toyota is a smaller machine with a wheelbase of 2,650 mm. and overall length of 4,655 mm., and weighs 2,395 kg. With its six-cylinder 3,878-c.c. engine producing 125 b.h.p. a top speed of 120 k.p.h. is possible.

Thornycroft Airfield Crash Tenders   *Plates 133–5*
With a World War II production of just over 5,000 the Thornycroft Nubian 4 × 4 chassis was a natural choice for use as a civilian vehicle for use in difficult conditions when maximum traction was required. The chassis was available with either a six-cylinder diesel or a four-cylinder petrol engine which drove through a transfer gearbox to both front and rear axles. Bodies for fire fighting use were supplied by such builders as Pyrene, Sun, Carmichael and Airfoam.

Later, a six-wheel version of the Nubian was introduced and soon

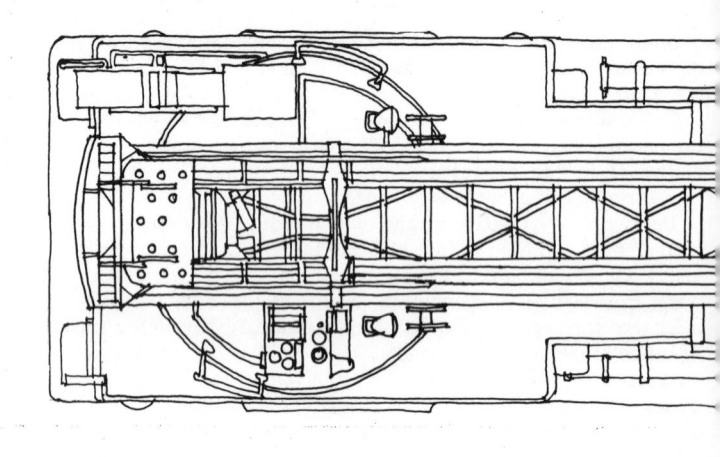

Fig. L Hino 32-metre aerial ladder, 1971 (Japan)

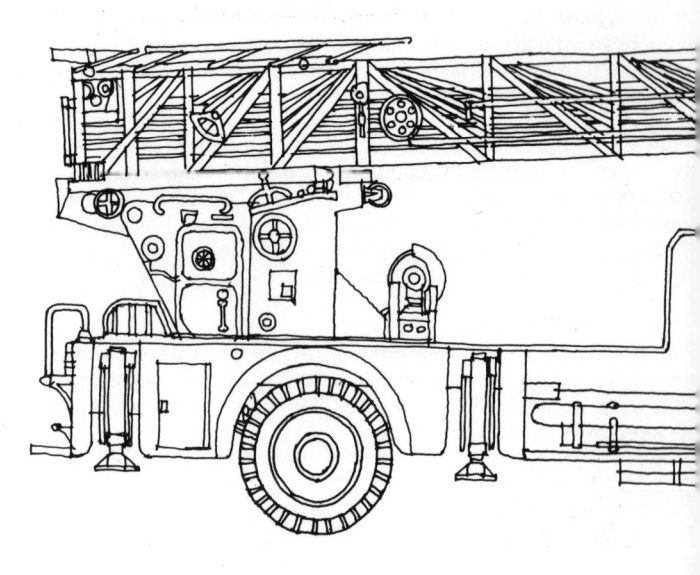

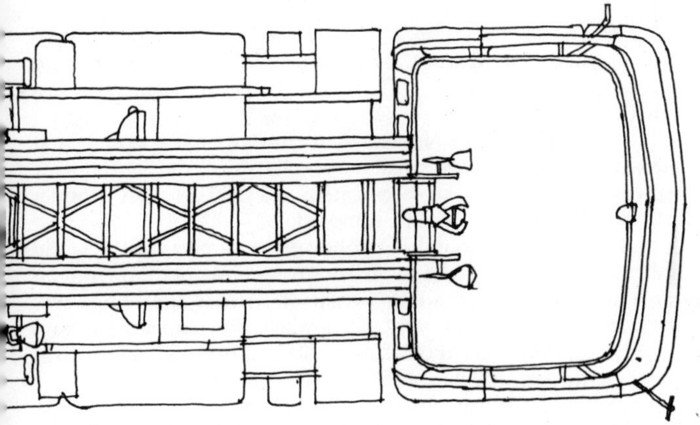

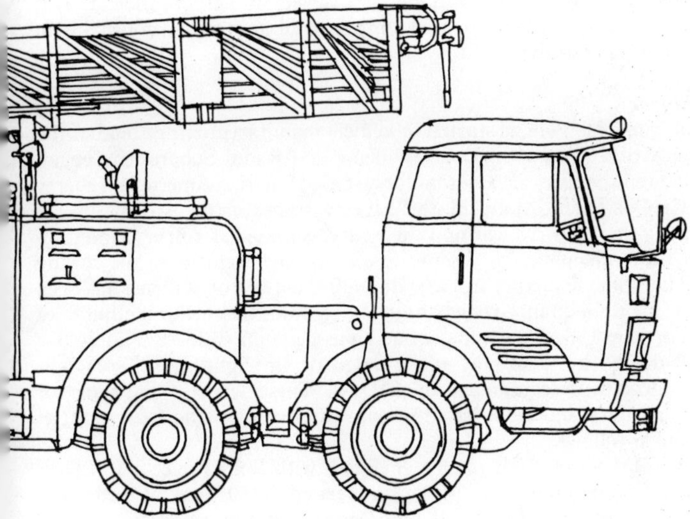

gained a reputation for being a most versatile go-anywhere chassis with its six-wheel drive feature. By this time the original Thornycroft six-sylinder oil engine was still offered but the petrol engine was now of Rolls Royce manufacture.

As the demand for the Nubian as a conventional goods vehicle declined and it became a fire/crash tender chassis, so the diesel engine option was dropped from the specification and the B81 eight-cylinder Rolls Royce petrol engine became standard.

Although the vehicle found many buyers in all parts of the world and it became a familiar sight on airfields from Finland to Australia it was obvious that with an all-up weight of 14 tons the engine was being called upon to work very hard when the tender went into action. Also aircraft were increasing in size and the attendant risks outstripping the performance of the best tenders.

The Nubian Major was designed to keep pace with the development of larger aircraft and provided the fire/crash tender designer with a chassis capable of grossing up to 20 tons. In order to meet the extra weight and still obtain a good performance the Cummins V8 300-b.h.p. diesel engine was installed and a five-speed semi-automatic gearbox in place of the synchromesh unit in previous models. Thornycroft approved fire/crash tenders have bodywork by Carmichael, Dennis, Gloster-Saro, HCB-Angus, Merryweather or Pyrene-Chubb and are available in a variety of configurations to suit the particular conditions prevailing at civil and military airfields in various parts of the world.

Willeme *Plate 136*

The name of Willeme started in vehicle manufacture at the end of the first World War when Louis Willeme and Raoul Schorestene began the manufacture of a 5-ton lorry based on the American Liberty design of truck. Many of the Liberty trucks were used in France during the 1914–18 war and their success provided a firm foundation for the Willeme design. From those early days right up to the closing of the Willeme factory in Paris in 1970 the name of Willeme has been linked with a quality vehicle built for rigorous tasks. For a number of years from 1962 the company had an association with AEC Limited.

During the 1960s a small range of fire-fighting vehicles was produced with equipment by SIDES. These were of the type for airport duty and were based on six-wheel chassis, the four varieties being as follows.

VIMP-AEC/Willeme 6 × 6 chassis with Rolls Royce 230 B.H.P. engine. Maximum speed 95 k.p.h. Carried 5,850 litres of water and 850 litres of foam compound. One large monitor on cab roof.

VITP-Willeme W8P 6 × 6 chassis with Willeme eight-cylinder 330-h.p. engine which gave a maximum speed of 90 k.p.h. Tanks contained 11,000 litres of water and 1,000 litres of foam compound. Vehicle had no monitors being designed as a foam tender to assist the monitor units in use.

VIGP (i) The Willeme W8P 6 × 6 chassis was again used as the basis of this appliance which, in addition to the storage of 9,100 litres of water and 1,100 litres of foam, also carried a single roof-mounted monitor and short extension ladders. Engine was the Willeme eight-cylinder 330-h.p. unit and a top speed of 90 k.p.h. was advertised.

VIGP (ii) A bonneted version of the Willeme 6 × 6 chassis W8P type but designated L8P. The same eight-cylinder 330-h.p. Willeme engine was listed as standard (type 518-T.8E). In addition to carrying considerable quantities of water and foam the vehicle had two large movable monitors mounted on the roof and four fixed low-level foam jets emitting through the front bumper. Facilities were also included for spraying foam on the ground beneath the vehicle. With an all-up weight of 27,500 kg. the vehicle was capable of reaching 60 k.p.h. in 1 minute with a top speed of 95 k.p.h.

Super Pumper Complex   *Plates 137–9*

When serious fires occur in some large cities it has been found that the main problem confronting the fire fighters has been that of obtaining an adequate water supply sufficiently close to the firescene. Another problem has been that of bringing additional water from hydrants or water supply points remote from the fire by means of relay pumping. This has proved inefficient and ineffective using conventional apparatus. Yet another problem has been that of having pumps powerful enough to hurl the vast quantities of water at the fire from a safe distance for the fireman.

This state of affairs had existed for many years, but only fireboats were capable of carrying the vast engines and powerful pumps which were necessary for the high-capacity, high-volume equipment needed. Another drawback on the use of such equipment was the difficulty of obtaining hose with the strength to withstand such high pressures and large volumes of water. The answer to the last problem mentioned lay in the research conducted for the United States Navy which produced high-pressure hose. Naval authorities were also instrumental in producing a powerful lightweight engine of the required output, for it was the British Admiralty which had promoted the development of the Napier Deltic engine as a power unit for boats. After the feasibility of using the Deltic engine and a powerful pump were studied by Gibbs & Cox, Mack Trucks Inc. decided to proceed with

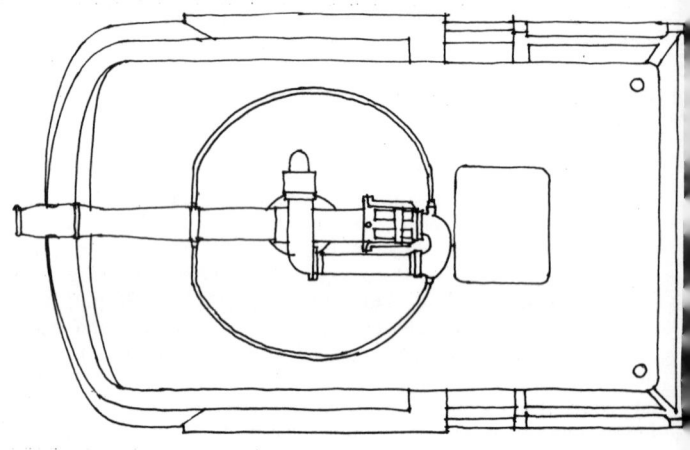

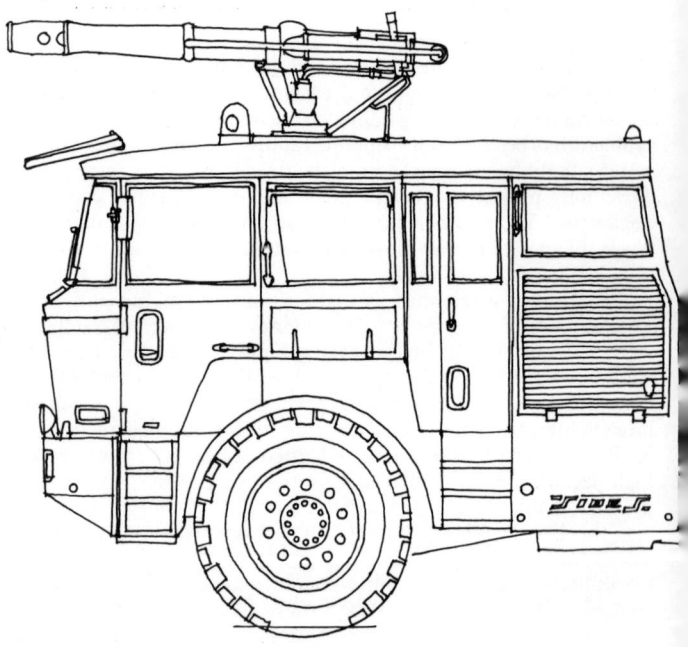

Fig. M Willeme-SIDES 6 × 6 airfield tender (France)

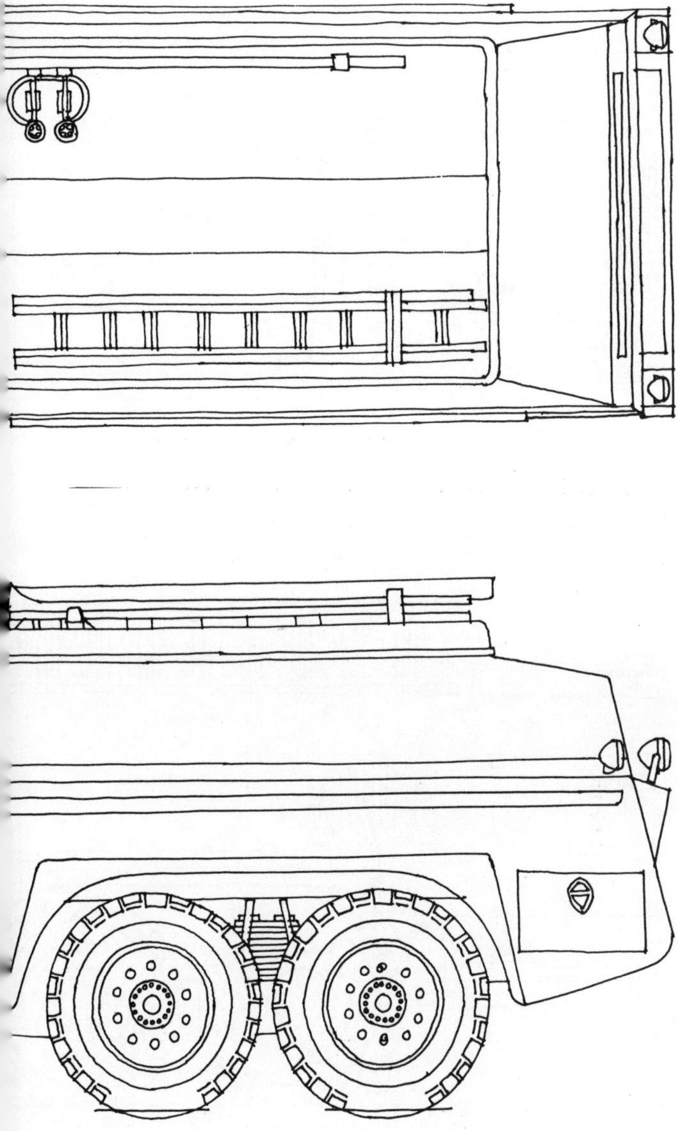

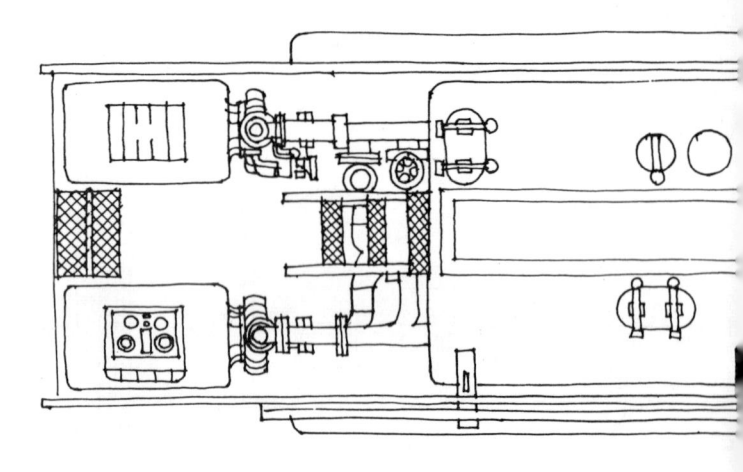

Fig. N Willeme-SIDES L8P 6 × 6 crash tender (France)

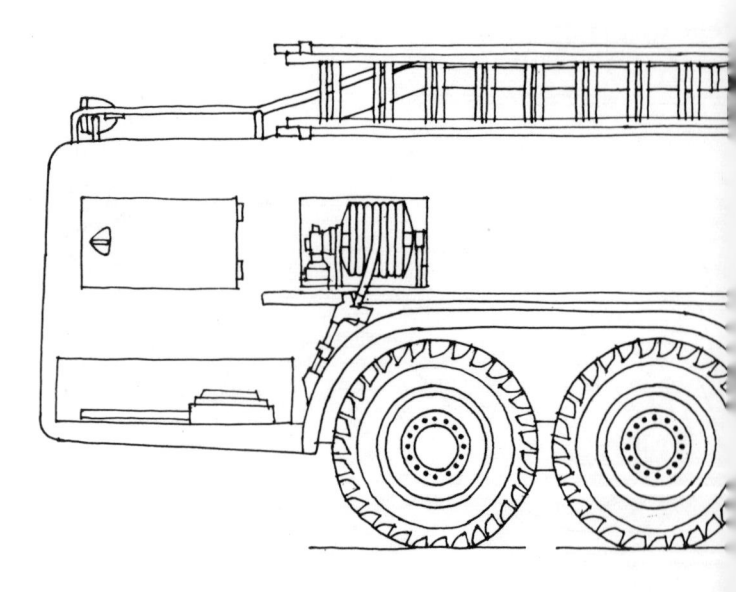

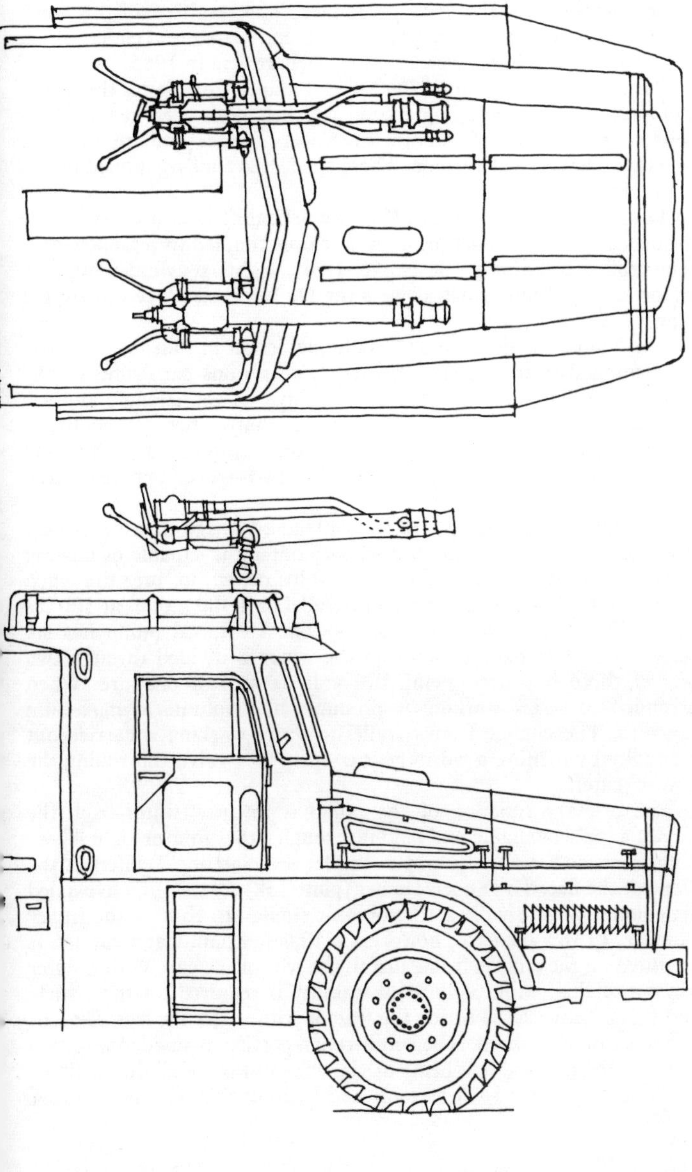

the development of the vehicle complex. The result of all this work was the Super Pumper, Super Tender and Satellite Tender which were built for the New York City Fire Department in 1965.

Basically the idea behind the Super Pumper Complex is the Super Pumper, a mobile pumping station which is positioned at the water source and is used to supply the Super Tender and up to three Satellite Tenders with water. These tenders are all positioned at the firescene.

The Super Pumper (plate 137) is an articulated outfit consisting of a Mack F715ST tandem-drive tractor unit coupled to a tandem-axle semi-trailer containing the Napier-Deltic eighteen-cylinder opposed piston diesel engine which powers the De Laval six-stage centrifugal pump.

The Napier-Deltic engine is of unique design in that the cylinders are arranged in triangular form with two pistons per cylinder. The three crankshafts are situated at each apex of the triangle and are geared together to drive a common output shaft. There are six banks of cylinders thus giving a total of eighteen cylinders and thirty-six pistons, the engine working on the two-stroke opposed-piston principle.

The lightweight engine is of such advanced design that it produces 2,400 h.p. at 1,800 r.p.m. and when pumping is capable of moving 8,800 gals of water per minute, at 350 lbs per sq. in. pressure when the De Laval pump is arranged in parallel or 4,400 g.p.m. at 700 lbs per sq. in. when arranged in series. The De Laval pump has six impellers and in parallel working the water is divided through two sets of three impellers giving full volume at half pressure. When arranged in series working it produces half volume at maximum pressure. The change from parallel to series working is carried out manually by turning a worm gear coupled to a valve controlling the flow of water.

The primary function of the pump is to draw water from the largest supply available and send it through either four or eight $4\frac{1}{2}$-in. high-pressure hoses to the Super Tender and Satellite Tenders at the scene of the fire. The Super Tender (plate 138) consists of a five-axled articulated vehicle with a tractor unit similar to that of the Super Pumper. In this instance, however, a 10,000-g.p.m. water cannon is mounted on the chassis frame just to the rear of the cab. With a water cannon of this output additional stability is required for the vehicle and this is achieved by fitting the tractor with hydraulic legs adjacent to the cannon position. The semi-trailer portion is divided into two sections, the forward half being used for the storage of all the ancillary equipment required such as tools and hydrant fittings, and the rear half to carry 2,000 ft of $4\frac{1}{2}$-in. hose. Under the hose compartment the four $4\frac{1}{2}$-in. water manifolds are led through a series of check valves and pressure reducing valves in order to reduce the incoming pressure of 700 lbs per sq. in. down to 100 lbs per sq. in. at the four $2\frac{1}{2}$-in. hose outlets at the side of the trailer. In operation the tractor and semi-trailer may be parted in order to fight the fire from two vantage points, the wheels of the semi-trailer being steerable from the rear of the trailer.

Up to three Satellite Tenders (plate 139) complete the Super Fire Fighter complex and these are of more conventional design, being based on four-wheel chassis and carrying a 4,000-g.p.m. water cannon fed by four $4\frac{1}{2}$-in. inlets. 2,000 ft of $4\frac{1}{2}$-in. hose is also carried on each of the Satellite Tenders.

With the Super complex consisting of Super Pumper, Super Tender and three Satellite Tenders all working together, up to 37 tons of water per minute can be thrown on to the heart of the fire.

Berliet GBK 18 *Plate 140*
The Berliet fire-fighting vehicle shown in plate 140 is a GBK 18 model four-wheel-drive chassis designed specifically for fire protection in forests. The robust 4 × 4 chassis is mounted on 10·00 × 20 tyres (twin rear) which are adequate to provide a high ground clearance for the vehicle. A six-cylinder petrol engine of 95-mm. bore and 110-mm. stroke with an output of 150 h.p. at 3,500 r.p.m. provides the 10,650 kg. vehicle with a top speed of 86 k.p.h. Drive is through a single-plate clutch, four-speed gearbox, two-speed transfer box and separate propeller shafts to both axles. Power-assisted steering is fitted, and braking is by air.

The main pump is situated at the rear of the vehicle and is of the centrifugal type P23 capable of moving 60 cubic metres of water per hour. Water can be taken from an outside source or from the 3,500-litre water tank and discharged through the two normal hose outlets or one high-pressure outlet or through the small-bore hose reel. An auxiliary centrifugal pump is mounted at the front of the water tank and is capable of pumping 6 cubic metres per hour. It is connected to the water tank and the hose reel and is driven by an independent 175-cc. two-stroke single-cylinder petrol engine producing 10 h.p. at 600 r.p.m.

The all-steel cab seats three and a bench seat over the auxiliary pump provides seats for a further four crew members. The hose reel is mounted high up across the rear of the water tank and the extension ladder is carried on top of the tank. Suction hoses are carried either side of the tank while delivery hose, axes, ropes, hydrant standpipes, nozzles, branches and portable extinguishers are housed in the side lockers.

The complete vehicle measures 5,705 mm. long, 2,450 mm. wide and 2,880 mm. high.

Water-borne Firemen   *Plate 141*

There are occasions when firemen need to take to the water in order to tackle a particular hazard or emergency. The one instance which springs to mind is that of burning buildings alongside a river, and for the protection of waterside structures fireboats are in use in many parts of the world. Fireboats also have the important task of being in readiness to tackle fires which occur onboard boats in river, dock and harbour areas.

An innovation of more recent times has been the use of small boats or rafts which can be carried on the roof of a specially-built appliance or towed behind an appliance on a special launching trailer. The function of these small craft is that they can be taken swiftly by road to a convenient point close to the emergency and then placed in the water by a small crane fitted to the appliance, or manhandled off the trailer. With some of the rafts used it is possible to negotiate quite shallow water areas which are inaccessible even by small boats. In some designs the portable pumping unit carried on the raft is used to propel the raft along by means of a water jet.

These small boats and rafts find use in dealing with many different emergencies other than fires, being able to traverse canals, lakes and flooded areas and carry out rescue operations. They do suffer the disadvantage of a limit on the amount of equipment they can carry and for powerful pumping duties one must resort to the true fireboat.

For a vehicle to achieve the compromise between a land fire engine and a powerful water-borne pumper we must turn to the amphibious fire engine, an example of which is shown in plate 141.

This particular vehicle was the first to be designed specifically as a fire engine and entered service with the Rheinland-Pfalz brigade of the Federal German Republic in 1968. With so many hazardous materials being carried in barges on the rivers Rhine, Mosel and Lahn, the Ministry of the Interior asked the firm of Eisenwerke (Kaiserslautern) to produce this unique machine in order to provide adequate protection for the area. The specification required that the resulting vehicle would be capable of 90 k.p.h. on the road and 50 k.p.h. over grassland, and be able to negotiate marshy riverbanks and open rivers and climb up the steep man-made embankments.

In use the vehicle has come up to its expectations. Power is supplied by an air-cooled twelve-cylinder diesel engine through a six-speed gearbox to both axles. For use in the water, a steerable propeller is provided at the rear of the machine.

Fire-fighting equipment consists of two 880-gals-per-min. centrifugal pumps, foam tank, dry powder tank, 9·5 kVA generator, air compressor, floodlights and radio.

Skoda ASC 16   *Plate 142*

A fairly recent (1962) type of appliance from Czechoslovakia is the Skoda ASC 16 type pictured in plate 142. This machine is based on the Skoda 706RTH chassis and is equipped with a rear-mounted single-stage centrifugal pump which produces 1,600 litres per minute at 8 atmospheres pressure. Water can be taken from either the 3,500-litre water tank mounted in the centre of the body or from a hydrant or open water supply. Six pump outlets are arranged at the rear. A 200-litre foam powder tank is situated near the pump and when connected, foam is produced at the rate of 3,000 litres per minute at 6 atmospheres pressure. In lockers adjacent to the pump are stowed lengths of flaked hose with permanently attached foam guns which can be brought into use immediately the vehicle arrives at the fire.

Two extension ladders are carried on top of the appliance together with a stretcher for injured personnel. Also on top of the body are the four lengths of hard suction hose carried in special tunnels. Access to these is by way of footholds and handrails at the rear of the appliance on either side of the pump hatch.

Large body lockers provide storage for all the other items of equipment which includes rolled delivery hose, hydrant fittings, ropes, axes and tools. The cab provides accommodation for the crew of eight.

Scania-Vabis Hydraulic Platform   *Plate 143*

In Northern Europe Sweden's two large commercial vehicle builders Scania and Volvo both produce fire appliances, although in small numbers. One type introduced in 1966 is the Scania-Vabis Hydraulic Platform shown in plate 143.

This appliance is based on the Scania LB 7650 chassis fitted with the Scania D 11 six-cylinder diesel engine which produces 190 h.p. at 2,200 r.p.m. from its capacity of 11 litres. With a wheelbase of 5,000 mm. and an overall length of 9.100 mm. the hydraulic platform closes down to give a height of 3,500 mm. and a maximum extended height of 235,000 mm.

The hydraulic platform and ancilliary fittings are supplied by K. Nummela Oy, Abo, Finland. Four outriggers are fitted to provide stability for the platform when in the working position and the boom can be rotated through a full 360° and extended to provide a horizontal reach of 13·8 m. Controls are placed on a console which forms part of the base fabrication of the boom and rotates with the boom. A duplicate set of controls is provided in the cage at the upper end of the boom.

A jointed steel pipe runs up the arms of the boom and terminates at the caged platform; sections of ladder are attached to the other side of the boom structure in order to provide access to the fabrication for repair or in cases of emergency. All movements are carried out by means of hydraulic rams powered by a pump driven by the vehicle engine.

Albion-Carmichael Firechief *Plate 144*

For many years before World War II Albion Motors built chassis for the mounting of special fire appliance bodies, notably by Merryweather and John Kerr. Since 1945 there have not been many fire engines using Albion chassis, although during the last ten years or so Albion Motors Ltd and Carmichael & Sons (Worcester) Ltd have together produced the Firechief as a water tender or pump escape (plate 144).

The Firechief chassis by Albion was specially developed for fire engine work from the highly successful Chieftain lorry chassis and at first was available in two wheelbase lengths, 11 ft 10½ in. and 13 ft 4½ in., both with the standard Leyland/Albion all-steel cab. Later versions had the wheelbase options of 12 ft 6 in. or 13 ft 6 in. and the chassis only was supplied to the bodybuilder, he supplying the full cab together with the rest of the bodywork.

The Firechief was powered by the Leyland L 400-S six-cylinder 6·54-litre diesel engine which produced 125 b.h.p. at 2,400 r.p.m. Drive was through a 14-in.-diameter plate clutch and six-speed manual gearbox, with a low 'crawler' and over-drive sixth speed, by means of open propeller shafts, to a heavy-duty rear axle of spiral-bevel type with hub reduction gear.

The Carmichael bodywork was of two main types, water tender and pump escape (or dual-purpose appliance), and both were available in composite construction employing steel or alloy metal parts. Finish was paintwork to customers' requirements or major alloy panels in self colour.

A full crew cab was provided which had seats for up to six personnel, and doors of the normal hinged type or folding jack-knife variety could be supplied as required. Water tank capacities from 50 to 1,000 gals were offered, the pump and hose reels were fully protected by the hinged doors or roller shutters which covered the lockers.

A Gwynne pump in three sizes was offered, namely, 600, 750 and 1,000 gals-per-min. output, or the Carmichael/Gwynne high/low-pressure pump could be supplied if required.

Bedford Fire Appliances   *Plate 145*

The Bedford truck was the successor to the Chevrolet in Britain and for a short while around 1931 was known as the Chevrolet-Bedford and Bedford Six. A number of fire engines had been produced by bodybuilders using the Chevrolet chassis and naturally the Bedford followed suit.

At first the short-wheelbase two-tonner was advertised as a complete fire engine for £740 but later, in 1933, a trio of appliances was advertised – the 30 cwt with pump, for £599, the two-ton with pump, for £730, and the 'special' for £850. This last type of vehicle used the bus chassis.

In 1935 the range included five different models and in addition there was the normal truck which was modified as a fire tender for the towing of trailer pumps or even old horse-drawn pumps. A saloon type emergency tender from this era is shown in plate 75.

One particular Bedford appliance supplied to Horsham brigade late in 1935 included in its equipment hay forks for dealing with fires in hay ricks, but more interesting than that was the fact that it was equipped with a two-tone siren – one of the first in Britain.

Just before the outbreak of World War II Bedford announced the Bedford-Martin Walter Utilecon as a light fire tender for towing a trailer-pump, and of course once the war was under way Bedford truck chassis became part of war production. Many Bedford chassis carried fire appliance bodywork during the war and they appeared as Home Office pumps in the heavy and extra-heavy class. Pumps were by Gwynne, Sulzer and Tangye with pumping engines by Leyland and Ford.

The post-war range of Bedford trucks included the K, M and O types, but not many appeared as fire appliances. When the Big Bedford seven-tonner was introduced it soon found buyers both at home and abroad and many hundreds were used as the basis for the pumps, control units, stores vans, pipe carriers, hose layers, foam tenders and kitchens of the Mobile Fire Columns in the post-war era.

The TK truck range has formed the basis of many fire appliances in the years since its introduction in 1960 and a water tender using the six-ton chassis is shown in plate 145.

Zuk A-14, 1968   *Plates 146–7*

In rural fire-fighting sometimes even the smallest light vehicles are unable to get close enough to the fire or the water source. If the outbreak is in a position totally inaccessible to a wheeled appliance of any sort then even a trailer pump can be rendered almost useless. It is in circumstances such as these that the portable fire pump really comes into its own. Of course the pump does not have to be removed from the carrying appliances – it will work perfectly well in either position.

For the swift deployment of such portable pumps a recent Polish design is the Zuk A-14 produced by Fabryka Samochodow Ciezarowych at Lublin. Designed to carry the pump together with the necessary 200 m. of delivery hose, two hard suctions, three metal ladders and a crew of five, the Zuk is provided with a four-cylinder 2,120-c.c. petrol engine of 82-mm. bore and 100-mm. stroke. The engine produces 77 h.p. at 4,000 r.p.m. and propels the loaded vehicle weighing 2,480 kg. at speeds up to 95 k.p.h. (Plate 146.)

The complete vehicle measures 4,406 mm. long, 1,820 mm. wide and 2,440 mm. high over ladders on roof. The demountable pump will deliver up to 800 litres per minute and is powered by a petrol engine of 26 h.p.

Also depicted is the A-15 model which is equipped with the M800E Typ PO3 pump driven by Typ S15M two-cylinder two-stroke petrol engine mounted within the rear of the body and tows a 290-m. hose reel mounted on a small two-wheel trailer. (Plate 147.)

Mowag Fire Appliances *Plates 148–9*

The two vehicles shown in plates 148 and 149 are from the small range of fire appliances produced by the Swiss Mowag concern at Kreuzlingen.

Only four-wheel vehicles are built, using their own 4 × 4 chassis or converted Chrysler equipment. Water tenders, pumpers, emergency tenders, rescue vehicles and pump/ladders are included in the present production.

The Chrysler 318–3 V8 petrol engine is used in the vehicles, and this power unit provides 210 b.h.p. from its 5,213 c.c. capacity.

The vehicle depicted in plate 148 is turned out in an unusual colour scheme in an effort to render it more conspicuous. There has been a great deal of effort and experimentation in recent years regarding the colour of emergency vehicles and the early wood finishes, reds, maroon- and green-painted appliances have given way to new colours and colour schemes.

There were of course many early appliances turned out in white, particularly in the United States and eastern countries, while at least one British horse-drawn fire engine was painted bright yellow. However, in many cases it was 'fire-engine red' for very many years and any variations stood out from the rest. Although reds and yellows are quite bright in daylight, the modern street lighting used in some towns can play tricks with colours and an engine that is red by day can appear black at night!

The recent appearance of reflective materials and fluorescent paints has helped the fire vehicle to be more easily noticed, and particularly in Germany the trend is toward vehicles painted fluorescent red or orange with distinctive white corner-panels and stripes.

The colour of fire apparatus is a problem that is still being debated and for the sake of this book we are glad they have appeared in such variety.

Mack Aerialscope *Plates 150–1*

The Mack Aerialscope is something of a cross between a turntable ladder and a snorkel. It is similar to a turntable ladder by virtue of the fact that it consists of a four-section extending boom with a fixed ladder on each section but the ladder is there for use only in time of breakdown of the boom. It is similar to a snorkel in that an operating cage is fixed to the head of the extending boom and used for fire fighting and rescue purposes.

The Mack Big Reach 75-ft CF Model Aerialscope consists of four rectangular-section metal tubes which are each of slightly different size, allowing the top three sections to be housed within the lower section when in the contracted or closed position. An operator's platform is attached to the top section of the boom and contains two monitors. The whole boom is mounted on a turntable base which has a duplicate set of controls to those in the operator's cage.

The boom is hydraulically elevated and extended and all the hydraulic, water and electrical equipment is located within the booms themselves. As noted above, each boom section has a length of ladder fixed to its upper side although this is provided for use only should the boom cease to function normally.

In operation the vehicle should ideally be placed parallel to the building and 32 ft from it. It is in this position that the Aerialscope can perform rescue operations to maximum advantage, covering an area of 6,850 sq. ft on the face of a building in a giant arc reaching up to 65 ft from ground level. The boom can also be depressed so that the cage is 10 ft below ground level. The boom can be rotated through a full circle and elevated to 75° from the horizontal.

Stability is achieved with the four vertical hydraulic jacks which are placed at the outer corners of the complete vehicle. In addition to these, two large hydraulic stabilizing jacks are angled out from the base of the turntable, one on each side of the vehicle. It is claimed that by means of these jacks the apparatus can be positioned close to a line of parked cars and the angled stabilizers extended in the narrow space left between the parked vehicles.

The carrying vehicle for the Aerialscope is a Mack cab-forward 218-in. wheelbase chassis with either a Mack six-cylinder o.h.v. gasoline engine (Thermodyne ENF 707C) producing 276 b.h.p. or a Mack turbo-charged six-cylinder diesel engine (ENDTF 673C Thermodyne) which produces 283 b.h.p.

Accommodation for the crew is provided by seats placed in front of and on either side of the centrally-mounted engine. Water connections are on either side of the body just below the turntable while a large covered locker to the rear provides space for the standard complement of eight ladders of various lengths which are carried on metal slides. Additional equipment is carried in side lockers, and the life net is in a separate locker at the rear of the machine.

USSR Appliances *Plates 152–3, 155*
The present range of Soviet made fire-fighting appliances available through the exporting agency of V/O Autoexport consists of seven types of vehicle plus a trailer pump and a demountable pump. The seven self-propelled units are based on three types of commercial vehicle chassis – the ГАЗ-66, a four-wheel-drive forward-control chassis, the ЗИЛ130 four-wheel normal truck chassis, and the heavy ЗИЛ 131 6 × 6 bonnetted truck chassis.

The ГА 3-66 chassis forms the basis for the АЦ 20 Fire Tank Truck which is basically a 1,610-litre water tank with a 55-litre foam agent tank alongside. The 1,200 litres per minute centrifugal pump is mounted at the rear of the vehicle below the tank and is driven by the vehicle engine via a p.t.o. It is capable of delivering water from the vehicle tank or from an open water source or hydrant. Facilities also exist for the mixing of the foam compound and water previous to discharge through the pump. Body lockers provide space for the storage of hydrant and other tools while the delivery hose is stored in rear lockers which when opened allow the hose to be fed out while the vehicle is on the move. Two lengths of hard section are carried in steel tubes carried above the water tank. For work in cold climates the АЦ-20 appliance has facilities for heating the cab and pump compartment and the water in the tank. An all-steel tilt cab provides accommodation for a crew of two.

Power is supplied by a V8 petrol engine which produces 130 b.h.p. and moves the vehicle at speeds up to 95 k.p.h. With a wheelbase of 3,306 mm and an overall length of 6,010 mm, the total all-up weight of the complete appliance is 5,820 kg.

The same four-wheel-drive chassis forms the basis for what is described as a Fire Communications and Lighting truck, type ACO-5, which is regarded as indispensible at large-scale fires and other calamities.

The standard all-steel tilt cab with accommodation for two crew members is backed up by a three-seater crew cab built into the vehicle body.

In order to fulfil its function as a communications vehicle the body contains two radio sets, an amplifier unit, a telephone set and automatic telephone exchange, six portable radios, a communications switchboard with jacks for connecting an outside microphone to the amplification unit, jacks for connecting extension loudspeakers and jacks for connecting the vehicle telephone into any city telephone system. Lighting functions are catered for by a vehicle engine driven generator, switchboard and three portable searchlights.

The Fire Pump Truck type AH-30 uses the bonneted ЗИЛ 130 truck chassis as its base, which is of the normal rear-wheel drive type. Powered by a V8 petrol engine which produces 170 b.h.p. the 8,000-kg.-weight vehicle is capable of speeds up to 95 k.p.h. on normal roads. A double crew cab provides accommodation for the crew of ten, side lockers contain hose, tools, standpipes and foaming agents, and the 1,800 litres per minute pump is contained in a compartment at the rear. A detachable hose reel is mounted at the rear of the vehicle and contains 120 m. of delivery hose. Extension ladders and suction hoses are carried on the roof.

Also utilising the ЗИЛ-130 truck chassis is the АЦ-30 Fire Tank truck which combines the use of water tender and fire pump. The vehicle has a double cab with space for a crew of seven, water tank of 2,100 litres and foam tank of 150-litre capacity. Side lockers contain most of the equipment which includes hose, axes, hooks, portable extinguishers, breathing apparatus and tools. The 1,800-litres-per-minute pump is mounted in a locker at the rear. A two-wheeled detachable hose reel containing six lengths of delivery hose is mounted at the rear of the body and must be removed to gain access to the pump. Suction hose, extension ladders, branchpipes and strainers are carried on the roof. For additional pumping power the MM-800 demountable pump of 800-litres-per-minute capacity can be supplied with the appliance if required.

To produce a machine with real cross-country ability the АЦ-40 Fire Tank Truck uses the six-wheel-drive ЗИЛН31 chassis fitted with the 170-b.h.p. V8 petrol engine used in the 130 chassis. Water and foam capacities of the appliance are 2,400 litres and 150 litres respectively and the contents of these tanks are discharged through the rear-mounted 2,400-litres-per-minute pump to normal hoses or to the roof-mounted monitor. The crew of seven are carried in the double cab and tools, hoses and equipment are stored in side lockers. As with the previous vehicle suction hose, extension ladders, branch pipes, and strainers are carried on the roof. The MM-800 portable pump can be carried on this appliance if required.

The 6 × 6 ЗИЛ-131 chassis is also used as a basis for the 30-m. turntable ladder rescue appliance, the АЛ-30 Fire Ladder Truck. One notable feature of this chassis is the ability it gives to vary the tyre pressures while the vehicle is in motion, this facility being controlled from the vehicle cab.

Power for the four-section ladder is provided by the vehicle engine driving hydraulic pumps via a p.t.o. The ladder can be rotated through a full 360°, extended by wire ropes to a height of 30 m. and projected up to an angle of 75° by means of twin hydraulic rams. In the event of power failure all movements can be carried out manually. Stability for the ladder is achieved by four angled hydraulic outriggers which relieve the vehicle chassis of any twisting associated with the weight of the extended ladder. The normal crew of three for the operation of the machine are carried in a single all-steel cab.

Ward La France Command Tower   *Plate 154*
Aerial ladders and hydraulic platforms are among the most expensive pieces of fire-fighting equipment, and yet there are times when even the smallest fire brigade could use an elevated position with advantage.

Realising that a demand existed for a low cost elevated tower of moderate height, the Ward La France Truck Corporation introduced the Command Tower to fill the need in small fire companies. The thinking behind the design is that even a small fire brigade with only one pump can have the Command Tower installed on existing equipment or specified as new equipment and so achieve certain advantages normally enjoyed only in large brigades with multiple apparatus. As the Command Tower is fitted on top of the machine just behind the cab it has little effect on the overall design of normal pumps. The 44 × 70-in. steel platform is raised hydraulically by means of twin three-section rams and supports two men. Access to the platform is by way of a three-section sliding ladder and the monitor is fed by means of a swivel-jointed 4-in. metal water pipe.

The Command Tower can be raised to a working height of 22 ft from the ground and affords its crew with a position for fire fighting, rescue or directing purposes. In plate 154 the equipment is shown mounted on a Ford chassis with Ward La France equipment and midships pump.

Faun 8 × 8 Airfield Crash Tender   *Plate 156*
With a history of fire apparatus building going back to 1845 the name of Faun has come to the fore recently in fire-fighting circles by virtue of their current range of vehicles. What is probably the most powerful and spectacular fire-fighting appliance in Europe today must be the Faun LF1410 8 × 8 Airfield Crash Tender with 1,000-h.p. engine (plate 156).

Specially designed to meet the hazards associated with the prospect of a jumbo-jet aircraft crashing during take-off or landing, this type of appliance was developed jointly by Frankfort Rhein-Main Airport, Faun, Metz and Total.

As viewed by the airport authorities the very real hazards created by a crashing Boeing 747 Jumbo-Jet aircraft is brought sharply into focus by just two facts: the fuselage can contain up to 490 passengers and the wing fuel tanks 180,000 litres of fuel. With an emergency situation such as this two main demands emerge: speed in reaching the aircraft and adequate power to quell the blaze upon arrival.

In order for the vehicle to be capable of carrying a high volume of fire quenching material, be it powder or liquid, it must be constructed in generous proportions. For this reason it must possess a high power-to-weight ratio to achieve an adequate performance.

It was decided that because of the high all-up weight of the complete machine coupled with the fact that an engine of 1,000 h.p. was desirable to maintain a high speed turnout, a layout with an eight-wheeled rigid configuration was necessary. A rigid layout was adopted not only because of its improved stability over an articulated vehicle, but also because it enabled all wheels to be driven, a most useful feature for an appliance which might have to traverse open ground at speed and be able to accelerate, manoeuvre and brake with equal alacrity. Unlike Great Britain, most European countries have seen no rigid eight-wheel vehicles save a handful for specialist applications such as crane chassis. However, Faun was an established producer of multi-wheeled chassis for a variety of cases and a rigid eight-wheeled goods-carrying vehicle had been produced in 1938.

The two variations of the LF1410 are a foam truck and a powder truck. Equipment for the foam truck is supplied by Carl Metz, while foam equipment for the second model is by Total Foerstrer & Company.

The Brief Specification of the Faun LF 1410/52v 8 × 8 Crash Tender is as follows:

1,000 h.p. (DIN) Daimler-Benz V10 diesel engine
ZF torque converter
ZF four-speed powershift transmission
Single tyres all round: 20·5-25 Michelin PR24
Gross Vehicle Weight approx. 53 metric tons
Acceleration: 0–80 k.p.h. in 45 seconds
Dimensions: 11,600 mm. long, 3,000 mm. wide, 3,120 mm. to top of cab
Capacities: Foam Tender, 18,000 litres of water and 2,000 litres of foam
Powder Tender, 1,200 kg. of dry powder

Pyrene Pathfinder   *Plate 157*

Introduced in 1971, the Pathfinder airfield crash tender is the joint production of Reynolds Boughton Limited and the Pyrene Company Limited.

Designed from the start as a high performance vehicle with great potential as a leader in the new generation of crash tenders, the Reynolds Boughton six-wheel drive chassis is powered by a General Motors V16 supercharged two-stroke diesel engine. With a capacity of 18·62 litres the rear-mounted engine produces 608 b.h.p. and drives by way of an Allison TC680 torque converter and six-speed automatic gearbox to the front axle and double drive rear bogie.

In order to afford maximum visibility to the driver he is positioned centrally in the spacious cabin, and just to his rear are seats for the other four members of the crew.

Fire-fighting equipment includes a 3,000-gal. water tank and 360-gal. liquid foam tank the contents of which are discharged by means of a Coventry Climax centrifugal pump capable of delivering 1,500 gals per min. The foam can be ejected from the remotely-controlled roof monitor which can throw a jet of expanded foam up to 250 ft at a rate of 13,500 g.p.m., or through the four sidelines which each produce 1,000 gals of expanded foam per minute. On either side of the vehicle is a normal water hose reel with 240 ft of small-bore hose.

The complete vehicle measures 38 ft long, 10 ft wide and 12 ft high, and weighs over 35 tons in fire-fighting trim. In spite of the great bulk of the vehicle the GM diesel engine can propel it at speeds up to 60 m.p.h. Acceleration is in the order of 0–50 m.p.h. in 34·4 seconds, and hill-climbing ability has been tested in an ascent of a muddy hill of 1 in 2·9 without resort to all-wheel drive.

Kronenburg Airfield Crash Tender   *Plate 158*

With a history going back to 1823 the fire-apparatus firm of

Kronenburg has a modern factory at Hedel which produces some of the best fire appliances in Europe. In addition to airfield crash trucks similar to that shown in plate 158 the company builds fire-fighting vehicles for city, urban and rural use. Trailer pumps, high-capacity foam tenders, police riot trucks and dry powder trailers all came within the scope of the company, and its products are exported to many countries outside the Netherlands. Many airfield crash trucks have been supplied to NATO airfields in Europe as well as to other military and civilian airfields.

The airfield crash truck depicted is the O-11-D type based on an FWD six-wheel-drive chassis incorporating two engines. A General Motors V8 two-stroke diesel engine is used to propel the vehicle while a GM V6 petrol engine provides power for the 2,000-litres-per-minute fire pump. By having two separate engines the vehicle is able to maintain a high speed while the pump engine is used at full capacity to produce up to 2,000 litres of foam per minute. In this way the vehicle can still provide its full striking power while being driven flat out toward the crashed aircraft or if desired lay a carpet of foam on the runway at high speed in preparation for the landing of a stricken aircraft.

A water tank of 5,800 litres and a foam tank of 1,200 litres are carried within the body of the vehicle and their contents can be discharged through roof monitor, side-mounted hand held hose reels or rear-mounted foam carpet nozzles.

Ward La France *Plate 159*

Ward La France are probably best known for the vehicles they supplied for the armed forces, although their reputation for building special heavy-duty vehicles, including fire apparatus, still ranks high among the major producers.

It was in 1918 that A. Ward La France started his vehicle building business at Elmira, New York. At first the production was sold in the immediate vicinity of the works but gradually as the products became better known vehicles were supplied to other states and later to other countries. The Second World War probably did more to spread the Ward La France name to other countries than anything else and some of the 5,000 wrecking trucks produced during that time are still in use.

The present range of fire apparatus consists of two series, the Constellation, which is Ward La France equipment on commercial chassis, and the Presidential series of custom apparatus.

Included in the Constellation series are the Courier brush truck, the Neptune tanker, the Comet pumper, the Venus aerial, the Saturn aircraft fire and rescue vehicle, the Aurora foam truck, the Apollo snorkel, the Gemini rescue vehicle, the Star lighting truck and the Command tower. (Plate 154.)

Featured in the Presidential series of custom apparatus are the Ambassador pumper, the Courier brush truck, the Diplomat aerial, the Senator snorkel, the Envoy tanker, the Delegate rescue vehicle, the Governor foam tender, the Congressman lighting truck and the Statesman aircraft fire and rescue vehicle. (Plate 159.)

Kaelble Airfield Tender   Plate 160
Although primarily concerned with the production of shovel loaders, crawler tractors, pipe layers and road rollers the German firm of Carl Kaelble does produce certain types of fire appliance chassis. One of these is the KV 700F 4 × 4 Airfield Crash Tender as illustrated in plate 160.

This machine uses a Daimler Benz MB 8V 331 diesel engine of V8 configuration employing turbo charging, which produced 710 h.p. at 2,200 r.p.m. from its 26·48 litres capacity. Drive is by way of an Allison torque converter and six-speed Power-Shift gearbox to the two Kaelble axles with locking differentials and carrying 26·5–25 single tyres.

Fully equipped the machine weighs 30 metric tons and has a top speed of 100 k.p.h. Acceleration is quoted as 0 to 80 k.p.h. in 40 seconds.

The machine's bodywork is by Kronenburg of Holland and includes centrally mounted tanks containing 8,000 litres of water and 1,000 litres of foam. The rear-mounted two-stage centrifugal pump has its own six-cylinder power unit which is mounted transversely at the rear end. The pump has an output of 2,500 litres a minute at 20 atmospheres and can supply the roof-mounted monitor, the two hose reel pistols and the under-truck nozzles. Facilities for refilling the foam and water tanks are grouped at the nearside of the vehicle.

Modern Firemen   Plate 161
In contrast to the firemen of the Insurance Brigades pictured in plate 16 with their picturesque livery and badge of office are the firemen of today (plate 161).

Important features of the old insurance brigade uniforms were that they were smart, distinctive and easily recognisable, whereas the modern fireman is clothed for service, function and protection.

For many years in Britain the fire brigades had drawn their recruits from among ex-naval men, and quite naturally many of the brigade uniforms had much in common with those of the navy. Some brigades did tend to overdress their men with uniforms designed more for appearance than protection, and in cases where the fire brigade was provided by the police force men often had to change uniforms when the alarm sounded!

Although it is the jacket and trousers one tends to think of when discussing uniform, it is the protection of the extremities of the human frame that have naturally received as much or more attention. The hands, feet and head and face are extremely important to the success of the fireman in his task, and it is in this area of the subject that most progress has been made.

In recent years the soft caps and brass helmets which have become legendary as the mark of the fireman have given way to more sophisticated types of headgear made of steel, cork and resin, reinforced plastics, glass fibre, etc. Some brigades still cling to the old styles of helmet with raised centre comb and lengthened back, while others are moving towards a smaller and closer-fitting helmet with additional face vizor and neck flap as required. For certain types of fire fighting such as in aircraft fires, a helmet which completely envelopes the wearer's head is favoured and these are supplied with face vizor and mesh guards as well as with an attached curtain to protect neck, shoulders and chest.

The helmet is an extremely important part of the protection of the fireman, but another item which plays just as important a part is the breathing apparatus which allows the fire fighter to get into close contact with the fire and yet not be overcome by the products of combustion. With a fire in an enclosed building smoke, fumes, dust, gases and burning particles may all be present in the atmosphere making breathing difficult. Filters may well be able to remove some of these hazards but the fire may have burnt up all the oxygen in the air so rendering the fireman's task impossible and making it necessary for him to retreat outside.

The first attempts at combating the hazards of breathing under these conditions involved providing a smoke mask, and this was put into practice during the eighteenth century. Some hundred years later an officer of the French Sapeurs Pompiers designed a hooded jacket which had an air pipe attached and was fed from outside the building.

The first self-contained breathing apparatus appeared at the beginning of the twentieth century and since that time has constantly been improved until today breathing sets are available embodying compressed air, oxygen or closed-circuit oxygen systems.

Featured in plate 161 are four modern firemen from Britain, Europe and America. The two British firemen are at the outside of the group and are shown at the left wearing breathing apparatus and at the right in normal fire-fighting garb. The Polish firemen at left centre sports a close-fitting helmet with low comb and small front peak, while the American fireman at right centre wears the familiar American-pattern helmet with large front and rear peaks.

FWD Aircraft Crash Tender *Plate 162*
In 1964 the FWD Corporation produced a spectacular aircraft fire-fighting and rescue truck of impressive design and performance. Although specially built for service with the US Air Force the P2 appliance was offered for civilian use with certain modifications.

Based on a rigid eight-wheel chassis with all wheels driven the vehicle incorporates four-wheel steering and is powered by two 340-h.p. petrol engines set side by side at the rear of the vehicle. The two engines are synchronised through a collection box in order to drive the vehicle at high speed across rough terrain. Although the complete vehicle grosses 65,000 lbs the power available can propel the vehicle from 0 to 55 m.p.h. in under 60 seconds. Top road speed of the vehicle is 65 m.p.h.

When required for fire fighting one engine drives through a p.t.o. from the collection box and drives the powerful 1,400-gals-per-min. centrifugal pump which in two minutes can smother an aircraft with foam from 200 ft away. The second engine is used to drive the vehicle while pumping is in operation, so rendering the apparatus capable of full fire-fighting ability while on the move.

Construction of the body is unusual in that it is built up of double-layer aluminium with foamed-in-place polyurethane, sandwich fashion, to form a 2-in.-thick wall. The 2,300-gal. water tank is similarly constructed while the 200-gal. foam compound tank is made of reinforced glass fibre.

In addition to the roof and front monitors a hand line with 150 ft of small-bore hose is fitted for dealing with small fires. A large cab accommodates the crew of four and is made of the same sandwich construction as the body. Overall measurements of the vehicle are 35 ft long, 8 ft 8 in. wide and 11 ft $6\frac{1}{2}$ in. high.

Ford Water Tender *Plate 163*
In both Great Britain and America the Ford chassis has been used by a wide variety of specialist bodybuilders as a basis for their fire apparatus.

The current Ford range is no exception. In Great Britain both the Transit range and the D series are currently being used by a number of bodybuilders to produce a variety of appliances in the small and medium weight range.

The Transit panel van with 175-in. wheelbase has formed the basis for small fire-fighting vehicles for rural or industrial use. Equipped with a 90-gal. water tank, rear-mounted pump, extension ladder and crew of four this machine with its lively V4 engine makes an ideal first aid machine.

By comparison the D series has been used as a basis for more appliances although mostly of the water tender type. HCB-Angus have produced Type B water tenders on D1013, D1014 and D1617 type chassis incorporating Turbo 360 diesel, 300-cubic-in. petrol or 165-b.h.p. V8 diesel engine.

Carmichael & Sons of Worcester also produce water tenders based on the 120-in. or 134-in. wheelbase chassis incorporating either diesel or petrol engines. A 400-gal. water tank is positioned over the rear axle. The Ford cab is suitably adapted to blend in with the rest of the bodywork yet still retains its facility of tilting, and a rear-mounted pump producing 500 gals per min. is fitted.

Of comparatively recent introduction is the Ford Water Tender incorporating bodywork by Jennings-ERF which again utilises the D1014 chassis.

Illustrated in plate 163 is a B-type water tender by Pyrene based on a Ford D600 chassis suitably modified by the special vehicle order department of the supplier. The 134-in wheelbase chassis is used and complete the vehicle measures 21 ft 2 in. long, 9 ft 1 in. high and 7 ft 6 in. wide. With so many vehicle chassis being produced at 8 ft wide the slightly narrower Ford has advantages in use by rural brigades where narrow lanes and gateways have to be negotiated.

Steyr-Rosenbauer, 1971 *Plate 164*

One of the modern productions of the old-established Rosenbauer company is shown in plate 164, the TLF 200 Tankloschfahrzeug on the Steyr 790 chassis. This is an interesting combination because it shows the complete Austrian fire-fighting vehicle, both Rosenbauer and Steyr being Austrian products, and both companies have recently celebrated 100 years of existence.

The Steyr 790 chassis has a wheelbase of 3,200 mm. and is propelled by a six-cylinder in-line diesel engine of 105 mm. bore and 115 mm. stroke which produces 170 h.p. All four wheels are driven and are shod with 900×20 14-ply single tyres.

The complete vehicle measures 6,750 mm. long and 2,480 mm. wide and is 3,000 mm. high, or 3,190 mm. including the cab-mounted monitor. Ready for the road the vehicle grosses 13,000 kg.

Utilizing the standard Steyr front end, Rosenbauer have produced a completely enclosed vehicle with extremely clean and pleasant lines. A water tank of 2,500-litres capacity is placed centrally in the body and the centrifugal pump is positioned to the rear of this being reached through a hinged door at the rear of the body. The Rosenbauer Type 65.000 pump is capable of producing 1,800 litres per minute and piping is provided so that the pump can supply the normal two outlets, the high pressure hose reel or the monitor, or a combination of these together. The monitor is controlled from within the crew cab, the operator standing on a flap-down stand which positions him head and shoulders through the opening flap in the cab roof.

With the exception of the suction hose which is stored under the body on either side, and the short ladders carried on the roof, all other equipment is carried in the easy-access side lockers which are reached through vertical rolling shutters. Rolled hose, portable extinguishers, hydrant fittings and other tools are all carried on special shelves in the side lockers. Two-tone horn, rotating blue lights and a detachable searchlight finish off the usual fitments.

Csepel-Ikarus 344, 1968 *Plate 165*

The Csepel and Ikarus concerns came into being in 1949 and 1948 respectively being the result of the nationalisation of the two Hungarian firms Manfred Weiss and Uhry Brothers. The Csepel half of the partnership produces the mechanical parts while Ikarus is the bodybuilding part.

The 344 type was first produced in 1968 being developed to meet the demand for a fully-equipped vehicle with cross-country characteristics whilst carrying the maximum of men and equipment within a maximum weight of 9,300 kg.

The chassis is of the four-wheel-drive type with twin rear tyres. A Csepel D 414h four-cylinder engine of 105 h.p. at 2,300 r.p.m. drives through a dry-plate clutch and five-speed-and-reverse gearbox to both axles. Both the clutch and steering have hydraulic assistance. A top speed of 85 k.p.h. is quoted with a cross-country speed of 40 k.p.h.

The bodywork design is commendable, being of aluminium framing with covering in polyester fibre. A water tank of 2,000 litres capacity is fabricated from 3-mm.-thick aluminium sheet, this being made possible by the special shape of the tank. Wide roller shutters on both sides of the body give access to the rolled hose lockers which provide storage for 640 m. of plastic hose. The centrally-mounted pump produces 1,600 litres per minute and has suction and delivery connections on both sides. A light-weight extension ladder is carried on the roof between the two tubes storing the suction hoses. The double jack-knife doors at each side of the body provide a rapid exit for the crew of twelve.

Magirus Deutz

Magirus Deutz have been producing fire equipment and vehicles since 1864 and today they are among the leaders in this specialised section of vehicle production.

Portable pumps, aerial ladders, pump water tenders, airfield crash tenders and special vehicles for dealing with specialised emergencies are included in the current range of equipment. The smallest standard item is a 250-gals-per-min. portable pump weighing 37 kg., whilst the largest vehicle is an eight-wheel-drive airfield crash tender of 1,000 h.p. carrying 18,000 litres of water and 2,000 litres of foam. Between these two extremes is a range of vehicles and equipment designed to combat all risks.

The vehicles illustrated in fig. O are of eight different types. The DL44h is a 19-tonne 44-m. turntable ladder with lift and space for crew of six. The DL30h has a turntable ladder incorporating a rescue cage, extended height 30 m., crew cab for six and hydraulic levelling jacks. The DL18/8h is the smallest of the motorised turntable ladders, and is available with manual or hydraulic operation. It extends to 18 m., weighs 6 tonnes and has accommodation for a crew of three. The LF16 is a pump tanker weighing 11 tonnes complete with space for a crew of nine and equipped with a pump producing 2,900 litres per minute at 80 m. head pressure.

The Tro LF 4000 is a dry-powder appliance with a capacity of 4,000 kg of extinguishing material. Powered by an engine producing 232 h.p. it is a machine designed for a high speed turnout.

With a lifting capacity of 20 tonnes and winching capacity of 15 tonnes the KW20 six-wheeler is a machine designed for heavy lifting tasks encountered by fire brigades. A 230-h.p. engine powers this 26-tonne crane.

Another vehicle which performs special tasks is the boat-carrying accident and rescue tender. The body contains a wide variety of cutting, lifting and rescue aids while a small boat with outboard motor is carried on the roof. A crane is mounted at the rear of the vehicle to lower the boat into the water and a special frame is provided to protect the cab from damage if the boat is manoeuvred round the front of the vehicle.

One of the larger vehicles in the range is the six-wheeled airfield crash tender type FLF40/10000 which completes the selection. Powered by a 640-h.p. engine this 30-ton appliance carries 9,000 litres of water and 1,000 litres of foam.

Rosenbauer

Some idea of the great variety of fire-fighting and ancillary equipment

Fig. O Magirus Deutz types from the 1972 range (Germany)

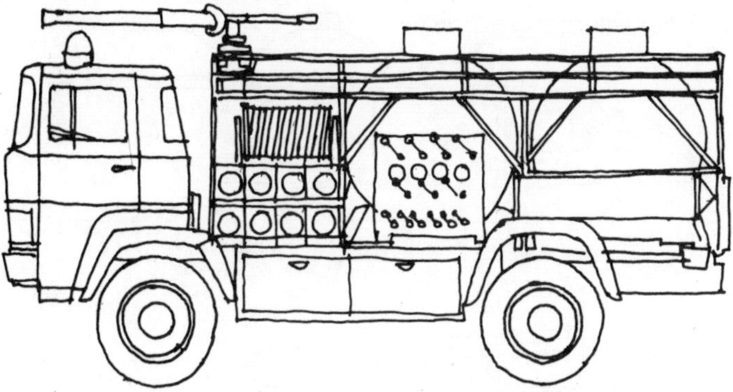

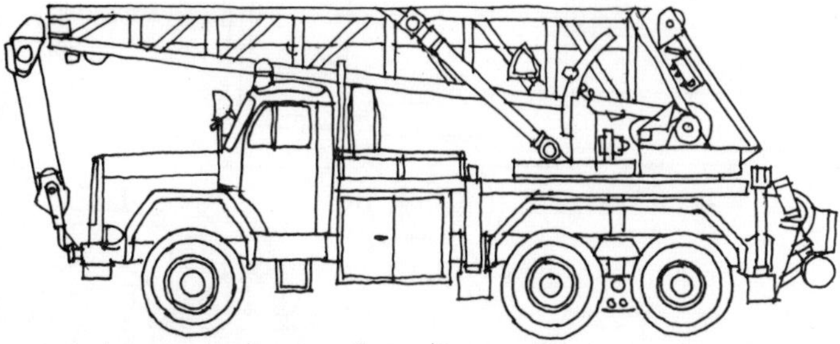

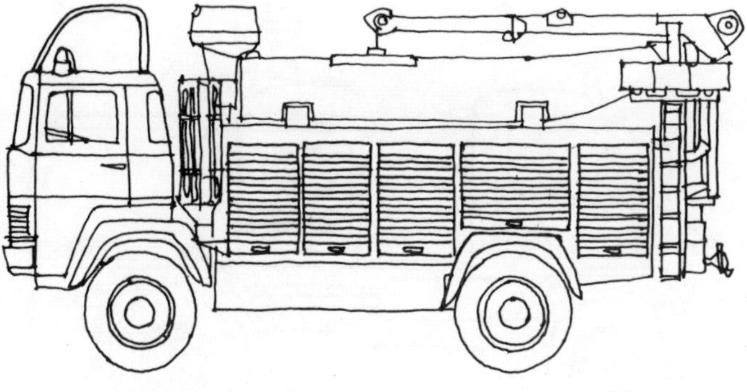

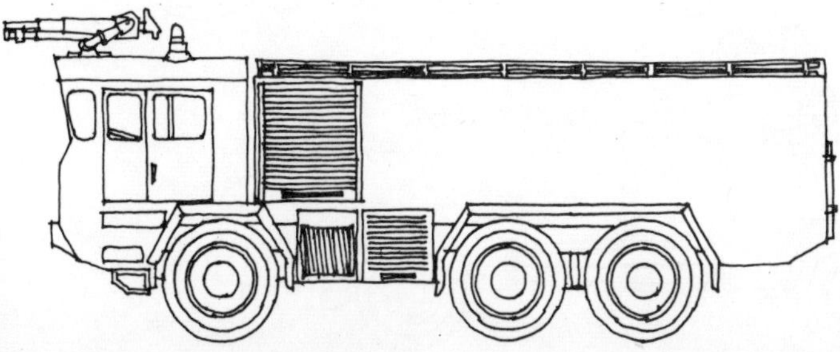

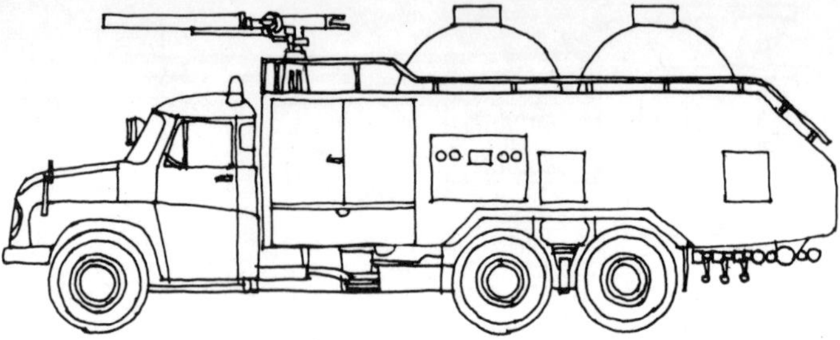

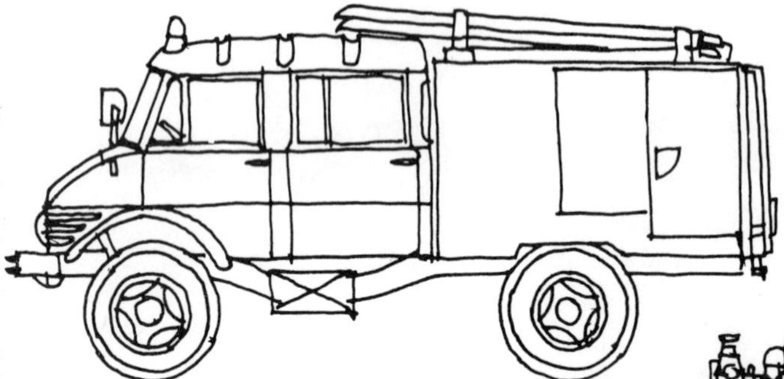

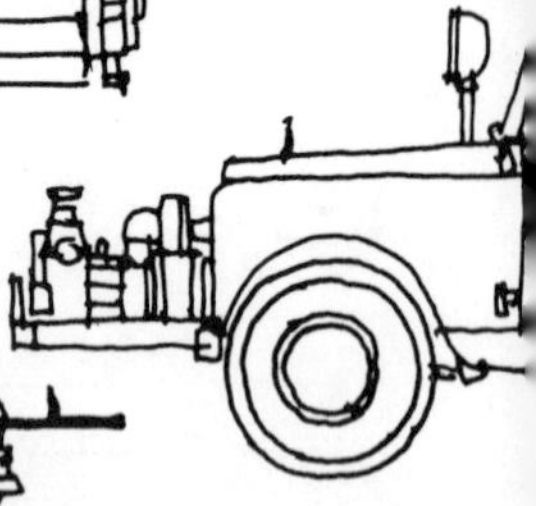

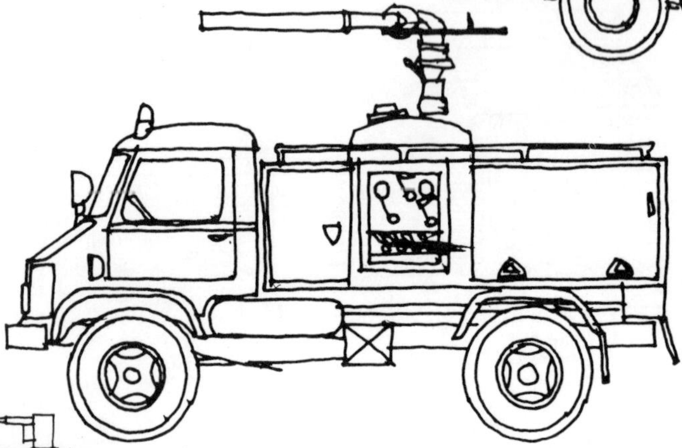

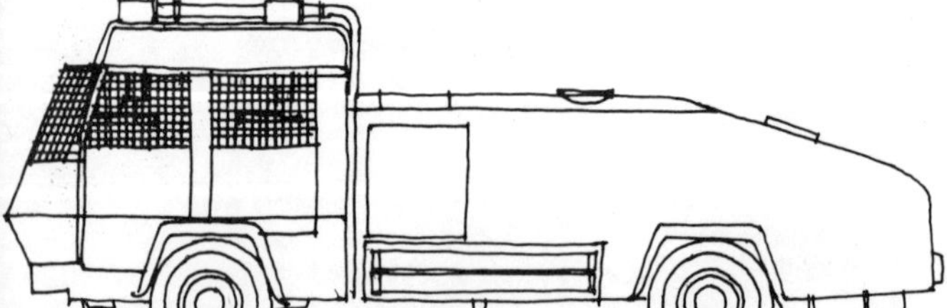

Fig. P Typical appliances in the Rosenbauer 1972 range (Austria)

produced by the Austrian fire engine works of Rosenbauer can be gained from the small display of varying types shown in fig. P.

The present range of products covers portable fire pumps, monitors, pump-water tenders, airfield crash tenders, dry powder tenders, foam tenders, water cannon, hydraulic platforms and breakdown tenders.

The size of machine available varies from small demountable pumps capable of being carried by two men to pumps or monitors mounted on trailers and vehicle-based equipment from Land-Rovers to large multi-axled crash tenders. Many types of vehicle chassis are used as the basis for Rosenbauer equipment including Land-Rover, Opel, Mercedes Benz, Kaelble, Faun, Dodge, Tatra, International, Perlini and of course the native Steyr.

American La France *Plate 166*

The origins of American La France go back to 1832 when John F. Rogers set up a business building manual fire engines in New York.

Through a number of amalgamations and sales agreements extending over many years the present company has emerged with a heritage which includes many names of importance in American fire fighting history.

The Button Fire Engine Works, the Silsby Manufacturing Company, the Ahrens Manufacturing Company and Clapp & Jones came together in 1891 to form the American Fire Engine Company. In 1900 the International Fire Engine Company was formed by the amalgamation of Thomas Manning Jun. & Company with the American Fire Engine Company and the La France Fire Engine Company.

A re-organisation in 1903 produced a new company, American La France Fire Engine Company Inc., and since that time there have been several other changes but the name American La France has continued.

The past history of the company has produced a wide variety of fire apparatus including manual, horse-drawn, steam- and gasoline-powered. Almost every type of fire equipment has been produced by American La France at one time or another including ladder trucks, hose wagons, water towers, aerial ladders, pumpers, chemical wagons, combination apparatus and articulateds.

The company has supplied complete apparatus or built on chassis supplied by other manufacturers. At one time normal truck chassis were produced but as the demand for fire equipment increased there was no room for truck building and they were discontinued in 1929 and merged with Republic trucks.

In 1909 the first gasoline-powered chemical wagon appeared using a specially-designed four-cylinder Simplex chassis. In these early days of gasoline-powered heavy vehicles fire chiefs still had all their faith in steam pumpers, and for a number of years there was business in supplying two-wheeled tractor units for replacing the horses and fore-carriages of horse-drawn equipment.

1914 saw the last steam pump built by American La France and in 1916 a new six-cylinder gasoline engine was produced further to stabilise the company's position with regard to motor apparatus. At this time three types of pumper were being built: a 750-gal. centrifugal pump, a 900-gal. piston pump and a 1,000-gal. gear pump. During the period 1910 to 1926 no fewer than 4,052 pumpers were turned out from the works.

Other famous names to join the American La France concern were Foamite Firefoam and O. J. Childs, who had joined together as Foamite-Childs in 1927.

A massive new 240-h.p. V12 engine was introduced in 1933, and when fitted into a pumper increased the output to 1,500 gals per minute. Toward the end of the 1930s two of these powerful V12 engines were used in a special Metropolitan Duplex pumper for Los Angeles. Two-stage pumps were used and an additional vehicle was required to handle the distribution of the water stream and direct it into twenty outlets.

Following upon the early vehicle type designations such as type 10, type 31, and type 75 came the 100 series which started in 1927. The 200 series followed in 1929 and the 300 series in 1932. 1934 saw the start of the 400 series followed by the 500 in 1938, the 600 in 1942, the 700 in 1945, the 800 in 1956, the 900 in 1958 and the 1,000 in 1970.

Several American La France apparatuses are shown in this book. Fig. G shows a motorised front-drive conversion of an old horse-drawn steamer. In plate 53 an old horse-drawn water tower is shown coupled to a 1925 Mack tractor, and in plate 108 is pictured a more modern water tower featuring two fixed monitors. More modern apparatus is shown in figs Q to V. These include a 1000 series custom Aerochief shown in both closed and extended positions (figs Q and R): a 1000 series custom pumper (fig. S); 1000 series front- and rear mounted aerial ladders (figs T and U); and a 1000 series custom Quint midships ladder (fig. V). In plate 166 is shown one of the Pioneer 11 range of budget-priced custom pumpers specially designed for small fire companies.

Commando   *Plate 167*
One of the phenomenal success stories of modern commercial vehicles

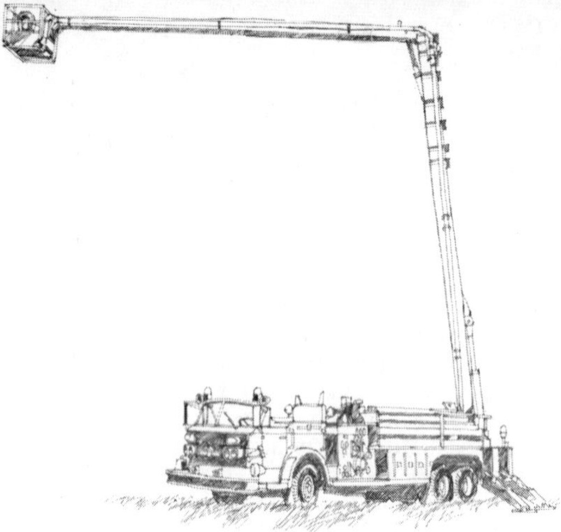

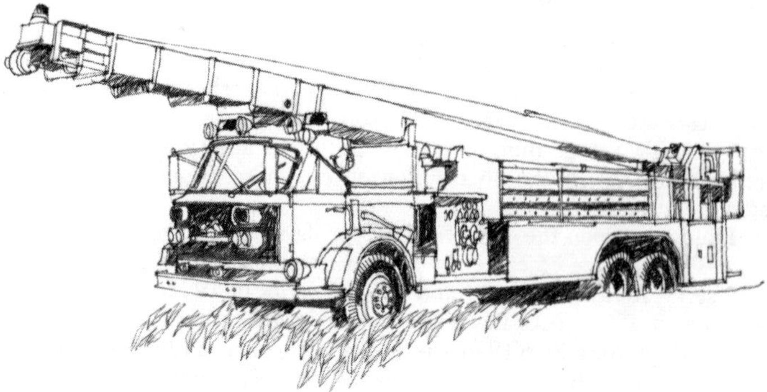

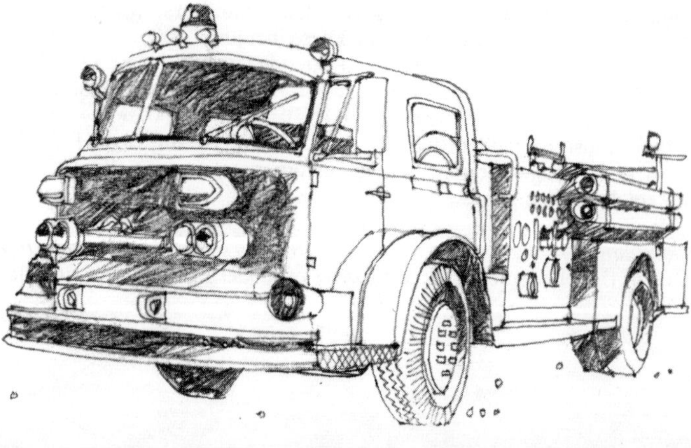

Figs Q and R American La France 1000 series Custom Aerochief, 1972, closed and extended (USA)

Fig. S American La France 1000 series custom pumper, 1972 (USA)

Fig. T American La France 1000 seri mounted aerial, 1972 (USA)

Fig. U American La France 1000 series rear-mounted aerial, 1972 (USA)

Fig. V American La France 1000 series quint aerial, 1972 (USA)

is that of the Land-Rover. Its luxury brother the Range Rover will almost certainly follow in the footsteps of its predecessor and in the short time since its introduction has already chalked up successes in several fields.

Another versatile vehicle has recently appeared in the shape of the Carmichael Commando, a six-wheel variation of the standard production 4 × 4 Range Rover with a training axle fitted to increase the space and payload (plate 167). The Commando is offered in three guises and brief descriptions are as follows:

Standard – within an overall length of 18 ft 6 in. is contained the normal Range Rover cab which together with additional space to the rear accommodates a crew of five. In the front of the bonnet a winch may be positioned, while to the rear of the cab is a van type body with side and rear rolling shutters giving access to the 155-cubic-ft locker space.

Water Tender – meeting Home Office Specification JCDD/18, the vehicle carries a 200-gal. water tank, front-mounted 500-gals-per-min. centrifugal pump, 180-ft hose reel at the rear and locker space in the van body. Provision can be made for carrying a ladder on the roof and if fitted this is mounted between the suction hose tunnels.

Emergency Aircraft Rescue – the unusual machine of the trio is this type designed as a first strike appliance for crashed aircraft. It is envisaged that with its rapid acceleration and 75 m.p.h. top speed this machine equipped with roof-monitor, pump, air compressor, water/foam tanks, inflatable jump slides and a crew of four can be quickly at the scene of an airfield emergency and render assistance promptly.

FIRE APPLIANCE DESIGN

Take any lorry chassis, couple-up a fire pump to the gearbox, add a ladder or two, put on some lockers for the hose and some seats for the crew, tie the lot together in a bright red body and you have a fire appliance – or do you?

No, it's not quite like that, not today anyway. True, at one time many fire appliances were merely adaptations or conversions of commercially available chassis, both private car and lorry, but today the fire appliance builder is faced with producing a complete machine which meets the demands of Chief Officers for a high-speed, expensive and highly sophisticated piece of fire-fighting equipment. This is not to say that the modern producer of this type of machine makes everything from the wheel nuts to the ladder-rungs. Today the production of a fire appliance may be the result of collaboration between several firms specialising in particular items of equipment.

Most builders of fire appliances have several 'standard' types which are produced to order and will be supplied as per published specification. Many are willing to vary their standard models to suit the requirements of a particular customer, providing of course that the variations requested are within the operating laws of the country where the vehicle is to operate and that the complete machine will not be so encumbered with additions as to make it unsuitable for the work envisaged.

In some countries the design of fire appliances may be dictated by a government department as in Great Britain where the Home Office issues a set of standards for fire-engine design, construction and performance. These standards do not place any imposition upon the appliance builder, but merely set out details of minimal conditions. If the appliance meets these conditions, the builder can quote the fact that his machine meets the requirements of 'JCDD — .' The JCDD number merely relates to the reference number of the design standards set out by the Joint Council for Design and Development of fire appliances set up by the Home Office and Scottish Office for Home Affairs.

Typical of the conditions set out in the JCDD are standards relating to, for the vehicle, weight distribution, road performance, chassis design, engine performance, fuel system, and electrical system, and for fire-fighting equipment, the pump, hose reels, water tank, controls, ladders, bodywork, lockers, crew compartment, etc.

The requirement specification is intended to state certain minimum requirements but there is nothing which prevents a purchaser from specifying that a vehicle is to have better requirements to suit his individual needs. For example should a Chief Officer decide that in order to meet the problem of maintaining good road speeds in a hilly district he will install an engine of greater capabilities than that detailed in the requirement specification, there is nothing to prevent him doing so.

So our professional fire engine builder in the shape of the chief engineer or draughtsman or designer is placed in an unenviable position when the actual production of a particular appliance is being contemplated. On the one hand it is the customer who will tell our designer the type of appliance he requires to be built and details the equipment and crew it is desired to carry. On the other hand are the limitations which encompass the project, whether they be physical, fiscal or judicial. Between these two extremes lies the designer.

Our designer may be given specific details of the particular chassis he is to use, the type and capacity of the pump he is to incorporate, the weight of all the loose equipment to be carried and other relevant details. He may be given precise instructions as to the entire layout in some cases, while in others he may be given broad limits to work within. In every case, he will require to know as much detail as possible in order that the finished product comes as close as possible to what was in the mind of the customer at the start.

It is probable that there will be lengthy discussions between the two parties in order to produce the machine required. It is often difficult to put onto the vehicle everything that is asked for and yet still be able to use the machine on almost every road in the area of operation and put up an acceptable road performance. It is all very well having a large, heavy machine which carries almost every item of equipment that will ever be required, but if it is so slow that it takes a very long time to reach the fire, or is so cumbersome that all but the main roads are inaccessible to it, then it will be of little use to a busy brigade. For reasons like this it is important that a lot of thought be given to the power/weight ratio and bulk of the machine in the early stages of design. To acquire a highly sophisticated and expensive machine only to find that in use it is slow, unstable or even downright awkward so as to be a liability on the brigade could be disastrous.

Let us turn our attention to fire-appliance design so far as the user is concerned and see how he tackles the problem, before returning to our designer and the problems that confront him.

Naturally there is no such thing as an ideal or best fire appliance for every fire risk. The type of machine which is needed for an area abounding with tall buildings and city streets would have limited use in an area which was rural in character with fire risks to match. In some cases the fire risk can be stated specifically and the machine to deal with that particular risk can be accurately detailed. Under this heading would come airfield crash tenders, foam tenders and hose layers where their sphere of operations can be closely controlled.

According to the risks involved it should be possible to detail the types of vehicle required to cover a particular area by means of one or more stations. In this way a detailed list of requirements can be drawn up showing gravity of certain risks, distances to be covered, availability of water supplies, terrain of area, accessibility of area, weight limitations, height restrictions and width of roads. Armed with such information the next step is to decide how best to carry the equipment needed to combat the risks of the area. Naturally, when a call is received by the brigade not every appliance will be sent out at once. There will probably be one particular machine labelled the 'first call' appliance and this would be sent to most calls when first received, although it may vary according to the risks associated with the premises to which the brigade is summoned. For instance the foam tender would not normally attend a grass fire nor would an amphibious vehicle be sent to a fire in an hotel.

Let us take a look at what could be the deployment of appliances and their equipment at a fire station in a large town, assuming that the area is a high-density one bristling with risks involving large buildings, homes, offices, industrial premises and perhaps a main trunk road not far away. The apparatus might consist of:

1. A powerful pump of 600 to 1,000 gals-per-min. output, together with first-aid water tank and hose reel, carrying 55-ft wheeled escape, two short ladders, portable extinguishers, set of breathing apparatus, supply of suction and delivery hose, hydrant fittings and tools.

2. A water tender with 400-gal. tank, 30-ft extension ladder, two short ladders, two sets of breathing apparatus, foam-making equipment, hose, branches, tools, etc., and a demountable pump of, say, 250 g.p.m. output.

3. A turntable ladder of 100 ft height with rescue lines, searchlight, hose and monitor for use as water tower, and a demountable pump of 250 g.p.m. output.

4. A emergency tender equipped with extension ladder, breathing apparatus, metal-cutting equipment, portable extinguishers,

demountable searchlights, first-aid tank and hose reel, first-aid pump for hose reel, stretcher, first-aid kit, resuscitation equipment, jacks, tools, ropes, etc.

In addition to the above, with the knowledge of any special risks in the area the Fire Officer may decide to carry certain special items peculiar to particular risks. For instance if the area abounded with canals, rivers or a large reservoir a small boat might be considered a wise investment for use in rescue and fire fighting. Another instance of special equipment is the rotating nozzles which are suspended in the hold of a blazing ship to quell an outbreak below deck and these might well be carried by a dockland station's appliances.

To take one more instance of designing the appliance to fit the particular need of an area we can look at a rural station where only one machine is kept. Here the need is for a comparatively small machine capable of reasonable speeds over the greater distances to be found in rural areas. The ideal machine should:

be capable of high speeds,

carry sufficient water to enable an attack to be made on a fire occurring in an area devoid of natural or piped water,

have a pump light enough to be carried by the crew to a position close to any water supply,

carry sufficient hose to enable water to be pumped from some point distant,

be so constructed that cross-country operation is possible, i.e. have high ground clearance and four-wheel drive.

With such a list of requirements the type of vehicle which emerges as suitable would be based on a production vehicle chassis having four-wheel drive and cross-country tyres on wheels large enough to provide the necessary ground clearance. The heaviest part of the equipment would be the large water tank which when full should contain about 400 gals. This amount should be enough to keep up fire-fighting operations while additional water supplies are secured if necessary. Extending ladders of 30 to 50 ft should be adequate to deal with fires and rescue from low-profile buildings normally associated with rural areas. Pumping power would best be provided by a trailer pump or demountable pump, although the ideal would be to have both a pump attached to the vehicle together with a demountable pump for manhandling to water source. A small-bore first-aid hose reel should be provided, and the longest amount of delivery hose and suction hose should be carried.

When only one machine covers a large area, and that machine is called to an outbreak of fire, the officer in charge has to decide if that one machine is going to be able to deal with the emergency. Should it be found that because of the size of the fire or the degree of the emergency the one machine cannot cope, then additional help will be summoned. For example a rescue from a tall silo might demand the use of a turntable ladder or hydraulic platform (snorkel), while a large grass fire on open heathland might require the assistance of additional pumps, greater water supplies or more hose from a hose layer.

It is against this background of a multitude of risks that the officer of a brigade must plan his fire-fighting force and deploy his equipment to the best advantage.

Let us now return to the original point of this chapter – the design of appliances – and study the position and layout of the various items which together make up the complete machine.

Vehicle Chassis

According to the type of body the chassis is to carry, the equipment and crew it will contain, and the use to which it will be put, details such as chassis length, spacing of axles, number of driven axles and size of wheels and tyres will be decided. The size and output of the engine, type of transmission, number of gears, drive-axle ratio and tyre size all have a bearing on the performance of the finished vehicle.

For instance a cross-country type of machine will require to be high off the ground and have adequate under-chassis clearance. It will also be necessary to have a strong chassis frame to withstand the flexing of the chassis over undulating ground. For similar reasons the vehicle should have all wheels driven for better adhesion, be shod with large cross-country-pattern tyres, and if possible have single tyres all round. The total all-up weight will decide the cross-section size of tyres to be used, and adequate clearance should be left around the wheels and springs.

Now to look at a different type of machine which in turn requires a chassis of different type – the turntable ladder. For about fifty years the turntable ladder was looked upon as the premier piece of fire-fighting equipment in large brigades, although in many instances it has lost pride of place to a more recent development in hydraulic platforms, the snorkel. However, both the turntable ladder and the snorkel have many things in common because they have a similar role to play in fire fighting and rescue.

The very essence of both types of machine is their accessibility to great heights, although the turntable ladder still has the edge on the snorkel with regard to attaining a greater height. Because of this reach requirement the stacking of the ladders or booms brings a problem with regard to storage. In order to make the overall height as low as possible for high-speed transport to the call, the carrying vehicle also requires to be as low as possible in order that the centre of gravity is within reasonable limits and low arches, garages and bridges do not prevent its use. So we need a long, low chassis frame for a start. The all-up weight of a long ladder can be in the region of 6 tons while an 85-ft snorkel will weigh half as much again, so axles, wheels and tyres must be of sufficient strength. As the weight is mounted so high off the ground there will have to be roadsprings and shock-absorbers strong enough to bear the sudden variations in load as the complete vehicle is swung round corners and roundabouts at considerable speed. Should the ladder be very long or the gross weight high it may be necessary to specify a six-wheeled chassis, while a very long rigid vehicle can promote handling problems in tight corners which would be best answered by placing the equipment on an articulated chassis.

Another point to be remembered with such a machine is the fact that most of the weight is borne by the rotating base section or turntable, so this must be centred over the rear axle of a rigid chassis or the swan-neck over the drive axle of an articulated outfit. Also, because of the design of the apparatus enabling it to be rotated through a full circle of 360°, it is necessary for some provision to be made which enables the weight when extended to be transferred direct to the road surface, and not taken through the vehicle chassis, springs, axles and tyres. This is attained by providing jacks placed near the turntable base to take the weight direct from ladder to road. With ladders this is usually by means of locking the position of road springs and axle in relation to the chassis, while with a snorkel the extending jacks may take the weight direct from base of turntable to road, missing out the vehicle chassis.

Pump

The positioning of this item has probably given rise to more discussion than any other problem in fire-fighting circles. For many years the usual pump position was at the rear of the machine in Great Britain and Europe while in America the front or midships position was favoured. Later the midships position gained popularity in Britain.

It has to be admitted that there are advantages and disadvantages for each position. Some brigades say that to have the pump up in front of the driver, whether the vehicle is normal-control or forward-control, is the best for positioning the vehicle near the water supply, especially at night, while others argue that the centre or midships position is most desirable because access to the pump and controls is from either or both sides, while others again feel that the rear position is ideal as it affords access from three sides of the vehicle.

Various disadvantages of the various positions are also brought up by Officers and crews. The front pump is more susceptible to damage, the rear pump requires a long drive-shaft, the midships pump is very inaccessible at time of breakdown. It is also easy to see that a rear-mounted pump is rather awkward to work with a wheeled escape in position, and is rather in the way with a machine having crew exit at the rear, although some machines have been produced with the New World body design having the rear-mounted pump almost under the floor.

The pump position is not only important with regard to maintenance and weight distribution but also from the point of view of its accessibility to the crew when actually at work at the fire.

Water Tank

On these appliances carrying a first-aid water tank this is usually only of small capacity, around 40 to 50 gals, and is placed close to the hose reel which it supplies, usually high up on the machine. On machines carrying larger water tanks such as water tenders or boosters the water tank when full is the heaviest part of the vehicle and must be positioned accordingly. In order that the complete vehicle should be as stable as possible even when cornering at speed, it is important that the water tank be mounted as low as possible and positioned so as to provide an evenly distributed load on the chassis frame. The tank should be mounted directly on the chassis, contain baffles to prevent surge, be constructed of non-corroding material, and be easily filled.

Ladders and Escapes

According to their length, ladders are normally carried on the roof of the appliance or hung on the side. Upon arrival at the fire scene ladders are often required just as urgently as the pump, so their accessibility is of paramount importance. Side-hung ladders are easily handled by the crew, but roof-stowed ladders require a short fixed ladder or foot-holds built into the body of the machine. The roof of the machine must also be suitably strengthened if heavy equipment is to be carried there.

A wheeled escape is carried at the extreme rear of the machine by means of positive mounting from which the escape can be readily dismounted. A form of clamp is also used at the head of the escape in order to prevent accidental dislodgement.

Hose

Because of its semi-rigidity, suction hose is difficult to stow on an appliance. On some early machines the suction hose was carried permanently attached to the pump and then curved round in a wide arc, the strainer end strapped to the body. On others it was carried in a U shape by taking it along one side of the machine, curving it over the front wing and bonnet and over the other front wing, finishing off on the opposite body side. In some designs the suctions have been carried in troughs, tunnels, boxes or racks. Sometimes it is carried at the side, under the floor, on the roof or inside the body.

Delivery hose is much easier to stow, although of course there is much more of it asking to be stored. The most widely used method is to roll delivery hose although some prefer to store it in folds or flaked. In some instances where large quantities of hose are carried such as in a hose layer it may be in long ready-connected lengths which are hung on special racks or hangers forming part of the vehicle body.

The most important thing to remember about hose is that it must be stored compactly and be readily to hand. Like a good house a fire appliance needs plenty of storage space and ample lockers for hose storage is essential. In fire brigade work the position of equipment in its respective locker is fundamental to the efficiency of the brigade in action. The crew are drilled to know exactly which locker contains which piece of equipment, and to suddenly vary the position, contents or means of opening any locker can be disastrous in an emergency. So the hose lockers should be within easy reach of the fireman on the ground, they should have wide-opening doors or flaps, and they should not have to be held open while the contents are retrieved.

With the hose-layer type of vehicle almost the entire body is given over to the storage of hose and it is usually stored in such a way as to render it easily discharged by one man standing at the rear as the vehicle is driven forward and the hose is payed out over the sloping tailboard. Careful re-packing of the hose after drying out is necessary to promote long life of the hose and to make it ready for the next call-out for high-speed hose-laying duties.

Crew Accommodation

The space set aside for the crew of an appliance will revolve around the type of appliance and the number of crew considered necessary. With an appliance of the turntable ladder pattern a crew of only two to four is carried so a large cab is inappropriate, whereas the pump or first-call appliance will require a cabin of ample proportions if it is to seat the eight or more carried on such a vehicle. As well as providing seating accommodation for the crew it is necessary to provide some elbow-room in order to provide for the usual procedure of allowing the crew to put on their uniforms as they are carried toward the fire. If carried on the machine the uniforms and equipment of the crew take up considerable room and additional equipment such as breathing apparatus will take up even more. Getting dressed in a fast-moving vehicle is not the easiest of things to do, so some form of padding is provided in the crew compartment to help out when a crash stop is made or centrifugal force takes over!

Also of importance is the prompt exit of the crew upon arrival at the fire scene so wide-opening doors are a must. Double-hinged, sliding, folding and jack-knife doors have all been tried on appliances at various times, but whatever system is decided upon the real essence of the problem is to disgorge the crew as swiftly as possible.

Although not every aspect of fire appliance design is covered in the foregoing an attempt has been made to show the reader how important is the pre-planning and attention to detail with regard to the modern fire appliance in the design stage. A modern machine represents a considerable capital investment, usually of the taxpayer's money, the fire officer and the designer both sharing the dutiful responsibility of producing a purposeful and viable fire appliance.

APPLIANCE TYPES

Although by no means exhaustive, the following list embraces short descriptions of the most widely used types of appliance. It will be appreciated that these descriptions are applicable to broad groups of appliances which come under the various headings according to the equipment they carry, and the purpose to which they are put. The names used by the various fire brigades in different parts of the world can vary, although they may relate to similar pieces of equipment. For instance, a mechanically-operated ladder which is fixed to a vehicle chassis but can be rotated through 360° would be described as a turntable ladder in Great Britain, whereas the same piece of apparatus in use in North America would be termed an aerial ladder. Similarly the turntable ladder could be just what its name states, or it might be equipped as a water tower without a pump. It might also have its own pump, tow a trailer pump, or be fitted with a demountable pump, but still come within the general term of a turntable ladder. So there is plenty of room for variety.

There are many appliance type names which are merely combinations of others, such as pumper and hose car, or pumper, chemical and hose car. With this kind of variation the possible permutation of type names is considerable.

Aerial Ladder
North American equivalent of the European turntable ladder, although, unlike its transatlantic counterpart can be mounted with the base at the front or rear of the vehicle. Aerial ladders are mounted on either rigid or articulated vehicle chassis.

Chemical
Chemical apparatus covers those appliances whose major role is fire extinction by some means other than just water. There have been many varieties of fire extinguishers based on chemical means rather than water. They include powders, gases and liquids. These machines usually have cylinders or tanks for the storage of their chemical agents, some being self-contained chemicals which act as extinguishers while others require that the chemical be mixed with water and/or air before being applied to the fire.

Combination
A machine which performs a combination of tasks or functions, such as a pumper, hose and ladder truck or a pumper, chemical and hose combination. Sometimes referred to as 'triples' or 'quads', according to the number of combinations adopted.

Control Unit
A covered vehicle consisting of accommodation for the Officer-in-Charge together with his personal staff, necessary for the control of the equipment and staff at the scene of a large fire, together with means for maintaining adequate communications with HQ and other units.

Crash Tender/Airfield Tender
An appliance designed to meet the special risks involved with crashed aircraft. Usually capable of high-speed approach, immediate attack and ample fire-fighting power.

Dual-purpose Machine
An appliance which has a dual role in a brigade – for instance a pump-escape, which performs the double duty of pumping water and carrying an escape.

Emergency Tender
According to the risks involved in certain areas, most brigades number among their appliances at least one such vehicle. As well as being equipped for dealing with emergencies at fires, this type of vehicle may carry such items as breathing apparatus, metal-cutting equipment, portable electric generators, floodlights, resuscitation equipment, power saws, lifting jacks and other forms of lifting tackle.

Escape Carrier
Name given to a vehicle constructed mainly for carrying a wheeled escape. In some instances additional ladders are carried together with other equipment such as hose, branches, hydrant fittings and tools.

Foam Tender
Fire appliance which carries considerable quantities of foam for fire extinction. Foam may be of the mechanical or chemical type and may be stored in large vessels for immediate application or small containers ready for mixing if in powder form.

Hose Layer
This is usually a covered vehicle in which the hose is stored in special racks or on hangers which enable the hose to be payed out of the vehicle and onto the ground as the vehicle is driven forward. The hose is normally carried in long lengths and may be ready-coupled for use. In some hose layers two lines of hose can be dispensed simultaneously.

Ladder Truck
A fire apparatus designed to carry a varied assortment of ladders – extending, hook, rescue, etc. In some cases referred to as hook and ladder trucks.

Pump or Pumper
The fundamental piece of equipment in any fire-fighting force. Name given to the vehicle which carries a pump at the front, midships or rear position, and invariably carries such other equipment as is necessary to enable the pump to function, i.e. suction hose, delivery hose, hydrant fittings and nozzles. Additional equipment such as water tank, hose reel, extension ladder, portable extinguishers and foam generators are also to be found on some pumps in service with brigades.

Pump-Escape
An appliance which carries a detachable wheeled escape in addition to the other items normally carried on a pump. Sometimes called a 'dual-purpose' machine because it serves both as a pump and an escape carrier.

Salvage Tender
After the fire has been put out there is a lot of work to be done in protecting buildings and their contents from further loss or damage. With this in mind the salvage tender is designed to carry the equipment required by the men alloted this task. Items will include tarpaulin sheets, ropes, weights, brooms, brushes, buckets, shovels, etc.

Snorkel
The fire-fighters' equivalent of the goose-necked crane or hydraulic platform. Of comparatively recent introduction, this appliance consists of a hinged boom in two or three sections according to size, arranged on a rotating turntable base with an operating cage at the extreme upper end. The boom can be rotated through a full 360° and

the operator's cage can be positioned almost anywhere within the area covered by the boom both in height and outward reach, from ground level to maximum extended length. Water is conveyed to the upper cage by means of jointed pipes carried either alongside or within the boom arms.

Trailer Pump

Water or foam pump mounted on a two- or four-wheeled trailer to be towed behind a fire appliance or other towing vehicle. The trailer pump may be used as a normal pump for fire-fighting purposes or, because of its great mobility may be positioned alongside the water supply in order to carry out boosting duty to other pumps nearer the fire. Some trailer pumps have the pump and power unit mounted on a quickly removable frame so that they may be detached for carrying by two or three men should the need arise.

Turntable Ladder

An extending metal ladder mounted on a turntable base and capable of being rotated through a full circle. It is used both for fire-fighting and rescue purposes, some types having a lift fitted for added safety. Turntable ladders are normally powered by the vehicle engine and are driven by mechanical or hydraulic means.

Water Tender

This type of appliance has its use in areas where the water supply may be inadequate for fire-fighting purposes. The vehicle carries a water tank of at least 400 gal. capacity and may also have its own pump, carry a demountable pump or tow a trailer pump. Ladders, hose, hose fittings and other tools may be carried according to requirements.

Water Tower

A rigid water pipe, terminating at its upper end in a nozzle, which is raised by mechanical or hydraulic means to the required height for fire fighting. The pipe is braced for rigidity and carried on a rigid or articulated vehicle chassis. In this design the operator remains at low level, the controls being placed on the vehicle.

Also the term used for describing a turntable ladder which has a monitor fixed at the head of the ladder.

BIBLIOGRAPHY

Listed below are some of the books and periodicals studied during the preparation of this volume, and which we recommend for further reading on the subject.

For those wishing to enjoy the membership of a club with facilities for the exchange of information etc., application should be made to one of the following:

The Commercial Vehicle and Road Transport Club
   (The Secretary, 50 The Firs, Daventry, Northants)
The Fire Brigade Society
   (The Secretary, 28 Fernwood Avenue, Streatham, London S.W.16)
The Historic Commercial Vehicle Club
   (The Secretary, 32 Acland Crescent, Denmark Hill, London S.E.5)

Periodicals
   The Fireman (ceased publication)
   Fire (originally *Fire Call*)
   Commercial Motor
   Motor Transport

Books
   A Practical Treatise on Outbreaks of Fire. S. G. Gamble. London, 1926.
   Fires, Fire Engines, and Fire Brigades. C. F. T. Young. London, 1866.
   London's Fire Brigades. W. E. Jackson. London, 1966.
   Fire Engines and Other Fire-Fighting Appliances. Science Museum. London, 1966.
   Fire-Fighting Vehicles 1840–1950. Olyslager Organisation. London, 1972.
   The Fire Service To-Day. F. Eyre and E. C. R. Hadfield. London, 1944.
   Manual of Fire Appliances for Mobile Fire Columns. Home Office. London, 1956.
   Manual of Firemanship, Part 2: Appliances. Home Office. London, 1965.
   Fire Pumps and Hydraulics. J. E. Potts and T. H. Harriss. London, 1942.
   Fire Engines. Glasgow Museum of Transport. Glasgow, 1971.
   100 Years of America's Fire-Fighting Apparatus. P. da Costa. New York, 1964.
   Fire Marks and Insurance Office Fire Brigades. B. Williams. London, 1927.

The Early Days of the Sun Fire Office. E. Baumer. London, 1910.
Historic Fires of the West. R. W. Andrews. Seattle, Washington, 1966.
Stories of the Fire Brigade. F. Mundell. London, 1895.
A History of the British Fire Service. G. V. Blackstone. London, 1957.
Our Fire Fighters. J. Anderson. London, 1944.
The Battle Against Fire. R. Brown and W. S. Thomson. London, 1966.
American Horse-Drawn Vehicles. J. D. Rittenhouse. New York, 1958.
This Was Trucking. R. F. Karolewitz. Seattle, Washington, 1966.
The Observer's Fighting Vehicles Directory, World War II. Olyslager Organisation. London, 1969.
The Book of the Fire Brigade. A. L. Haydon. London, 1913.
Fire and the Fire Service. G. A. Perry. London, 1972.

INDEX

Figures and letters in bold type refer to the colour plates and text drawings; other references are to page numbers.

Adler 167
AEC **129,** 20, 177, 182
Ahrens-Fox **69, 70, 71,** 147
Airfoam 179
AJS 140
Albion **144,** 158, 175, 192
Alvis **119,** 151, 171
American La France **48, 108, 166, G, Q, R, S, T, U, V,** 136, 212
Argyll 13
Aster 135
Austin 20, 160
Austro-Daimler 131, 146
Austro-Fiat **65,** 142, 146
АЛ 30 **155,** 197, 198
АЦ 20 **152,** 196
АЦ 40 **153,** 197

Baddeley 12
Barbon 119
Bedford **75, 109, 113, 114, 117, 145,** 20, 151, 167, 168, 170, 193
Belle Isle 142
Belsize 13
Berliet **140,** 142, 167, 189
Boyer **57,** 141
Braithwaite and Ericsson **24, 25,** 12, 124
Braun **40, 41, 42, 45,** 132, 133
BSA **56,** 140
Buick 18
Bussing 142, 167
Button 212

Carmichael **120, 144, 167,** 179, 182, 192, 213

Cedes **38,** 131
Chambers 18
Chevrolet 141, 193
Childs and Webb 137
Christie 136
Chrysler **148,** 194
Citroen 167
Clydesdale **57,** 141
Commer **59,** 20, 142, 159, 167
County 160
Coventry Climax 162, 169, 200
Cross 136
Crossley **80, 82,** 152
Csepel **165,** 206
Ctesibius 12

DAAG 167
DAF 167
Daimler 18
Delage **79**
Delahaye **60,** 143
De Laval 188
Dependable 141
Dennis **94, 95, 96, 97, 98, 103, 104, 126, 127, 128, B,** 13, 19, 155, 156, 158, 166, 175, 182
Dodge **112,** 20, 141, 169, 212
Diamond T 141
Durkopp 167

Eisenwerke **141,** 190
ERF 20, 167, 205

Faun **156,** 167, 199, 212
FBW 167
Federal 141
Foamite-Childs 213

233

Fiat **73,** 142, 167
Ford **51, 52, 63, 74, 81, 84, 85, 100, 101, 102, 105, 107, 108, 163,** 20, 138, 141, 145, 150, 152, 153, 160, 161, 163, 167, 168, 204
FWD **158, 162,** 201, 204

Gamer 154
General 137
Gloster Saro 182
Goliath 159
GMC 18, 141, 200, 201
Guy 154
Gwynne 193

Hadley and Simpkin 121, 129
Halley **A,** 13
Hansa Lloyd 167
Hatfield **76,** 134, 135, 151, 162
HCB-Angus **121,** 172, 182
Hino **130, L,** 178
Hodge **26,** 12, 125
Howe **52, 63,** 137, 138, 145
Humber **106,** 163

Ikarus **165,** 206
International **124,** 141, 173, 212
Isuzu **131,** 179

JAP **103,** 142
Jeep **86,** 153

Kaelble **160,** 167, 202, 212
Karrier **83,** 153
Keeling 121
Kenosha 137
Kerr 192
Kleiber 136
Knox 137
Kronenburg **158, 160,** 200, 202
Krupp **64,** 142, 146

Laffly **61,** 143
Land-Rover **120, 121,** 20, 171, 212, 216
Latta 125
Laurin and Klement **46,** 134
Lay 125
Leyland **56, 89, 91, 125, I,** 13, 18, 140, 154, 156, 157, 158, 163, 166, 167, 174, 176, 192
Locomobile 137
Lohner Porsche 130
Lott 129
Lucar **2,** 118
Luitmeiler 137
Luverne 137

Mack **53, 54, 55, 76, 137, 138, 139, 150, 151,** 136, 139, 141, 148, 183, 195
Mag 142, 147
Magirus Deutz **O,** 207
Magomobil **68,** 147
MAN 146, 167
Maudslay 20, 178
Mavag 142
Maxim **122, 123, 124, K,** 137, 172
McLaughlin **50,** 138, 151
Meray **58,** 142
Mercedes Benz **87, J,** 131, 153, 166, 212
Merryweather **28, 29, 36, 37, 47, 76, 93, 129,** 12, 13, 18, 20, 126, 129, 134, 140, 151, 156, 158, 162, 177, 182, 192
Metz **128, J,** 163, 176, 199
Meyer Hagen 159
Minerva 142
Moreland 137
Morris Commercial **77, 111,** 152, 154, 169
Mowag **148, 149,** 194

NAG 167

Napier 18, 183
Newsham **18, D,** 12, 121
Nummela 191

Obenchain and Boyer **57,** 140
Opel 167, 212

Packard 18, 141
Perkins 177
Perl 146
Perlini 212
Pirsch **31, 90, E,** 126, 128
Pittler 133
Poole and Hunt **27,** 125
Pope Hartford 137
Pyrene **109, 110, 111, 112, 113, 114, 115, 116, 117, 118, 157, 163,** 168, 169, 171, 176, 179, 182, 205

Raba **64, 65,** 142, 145
Range Rover **167,** 149, 216
Reo **H,** 150
Republic 212
Reynolds Boughton **157,** 200
Roberts **30,** 12, 126, 129, 131
Robinson 137
Rolls Royce **72,** 171, 176, 177, 182
Rosenbauer **66, 67, 164, P,** 146, 205, 207

Saurer 167
Saviem 167
Scammell **99, 115,** 120, 149, 159
Scania-Vabis **49, 143,** 137, 167, 191
Scemia **62,** 144
Seagrave **F,** 135, 136, 172
Selden 18
Sellars and Pennock 11
Shand Mason 12, 125
Shawk 125
Shelkton 126

Silsby 212
Simonis **74, 88,** 131, 150, 154
Simplex 213
Skoda **142,** 134, 191
Somua 142
Star 18
Sternberg 'es Kalman **33,** 128
Steyr **66, 67, 164,** 146, 167, 205, 212
Studebaker 18, 141
Sulzer 161
Sun 179

Tatra 167, 212
Tempo 160
Thornycroft **88, 133, 134, 135,** 151, 154, 155, 177, 179
Tilling Stevens 13, 161
Total 199
Toyota **132,** 179
Triangel **92,** 156
Triumph **78,** 152
Tuedloff-Dittrich **34, 58, 68,** 128, 141, 147

Unic 167
Universal 137

Van der Heyden 11, 121
Vespa **110,** 169
Volvo 167, 191
Vulcan 18

WAF 146
Wagon und Maschinenfabrik **39,** 132
Ward La France **154, 159,** 198, 201
Waterous **137,** 145
White 141
Willeme **136, M, N,** 182
Winn 151
Wivell 124
Wolseley **35,** 129

Zuk **146, 147,** 193